Design of Steel Structures: Principles and Practice

Design of Steel Structures: Principles and Practice

Adalberto West

WILLFORD PRESS

www.willfordpress.com

Published by Willford Press,
118-35 Queens Blvd., Suite 400,
Forest Hills, NY 11375, USA

ISBN: 978-1-64728-498-5

Cataloging-in-Publication Data

Design of steel structures : principles and practice / Adalberto West.
p. cm.
Includes bibliographical references and index.
ISBN 978-1-64728-498-5
1. Building, Iron and steel. 2. Steel, Structural--Design and construction. 3. Structural design.
4. Structural analysis (Engineering). I. West, Adalberto.

TA684 .D47 2023
624.182 1--dc23

For information on all Willford Press publications
visit our website at www.willfordpress.com

Willford Press

Contents

Preface

This book aims to highlight the current researches and provides a platform to further the scope of innovations in this area. This book is a product of the combined efforts of many researchers and scientists, after going through thorough studies and analysis from different parts of the world. The objective of this book is to provide the readers with the latest information of the field.

Steel is widely used in the construction of large column free structures owing to its property of high strength per unit mass. Steel structures are commonly used in the design of water tanks, chimneys, transmission towers for microwave and electric power, roof trusses for factories, cinema halls, auditoriums, etc. Some other applications of steel structures include plate girder and truss bridges for railways and roads, and crane girders, columns and trussed bents in industrial structures. There are various advantages of steels over other materials for purpose of construction. These include high durability, assured quality, reusability, water and gas resistivity, and speed of construction. Another important benefit of steel structures is that they can be strengthened any time later through the process of welding. This book outlines the principles and practice for the design of steel structures. It includes some of the vital pieces of work being conducted by experts across the world, on various topics related to design of steel structures.

I would like to express my sincere thanks to the authors for their dedicated efforts in the completion of this book. I acknowledge the efforts of the publisher for providing constant support. Lastly, I would like to thank my family for their support in all academic endeavors.

Adalberto West

1

Introduction to Design of Steel Structures

1.1 Properties of Steel

Physical Properties

Physical properties of structural steel, as detailed by cl.2.2.4.1 of IS 800:2007, irrespective of its grade may be taken as:

- Unit mass of steel, $p = 7850\ kg/m^3$.
- Modulus of elasticity, $E = 2.0 \times 10^5\ N/mm^2\ (MPa)$.
- Poisson ratio, $\mu = 0.3$.
- Modulus of rigidity, $G = 0.769 \times 105\ N/mm^2\ (MPa)$.
- Coefficient of thermal expansion cx. $= 12 \times 10^{-6}\ /^\circ C$.

Properties of Structural Steel

The following are the factor to be considered in mechanical properties of structural steel:

- Chemical composition.
- Rolling method.
- Rolling thickness.
- Heat treatment process.
- Stress history.

The important mechanical properties of steel are ultimate tensile strength, yield or proof stress, ductility, weld ability, toughness, corrosion resistance and machinability.

Ultimate Strength or Tensile Strength

Ultimate strength which is the minimum guaranteed ultimate tensile strength (UTS) at which the steel would fail is obtained from a tensile test on a standard specimen, generally called a coupon. A typical specimen as per IS 1608 is shown in Figure. In this test the gauge length Lg and the initial cross-sectional area A_0 are important parameters. The dimensions of the specimens are established to ensure that failure occurs within the designated gauge length. The test coupons are actually cut out from a specified portion of the member for which the tensile strength is required. The initial gauge length is taken as 5.65 A_0 in the case of a specimen with a rectangular cross section and five times the diameter in the case of a circular specimen.

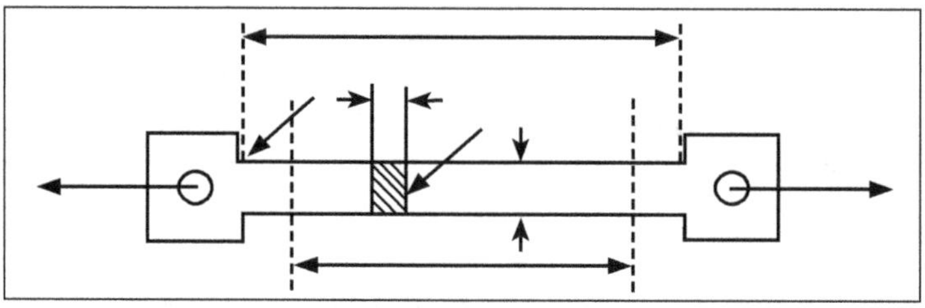

Standard tensile test specimen as per IS 1608.

The coupon is fixed in a tensile testing machine with specified distances between the grips and tested under uniaxial tension. The loads are applied through the threaded ends. A typical stress–strain curve of ordinary and high-strength steel specimen subjected to a gradually increasing tensile load is shown in Figure (a) and the stress–strain curve of mild steel specimen is shown in Figure (b).

The ultimate tensile strength is the highest stress at which a tensile specimen fails by fracture and is given by:

$$\text{Ultimate tensile strength} = \frac{\text{Ultimate tensile load}}{\text{Original area of cross section}}$$

Note that steel is specified in the code by the characteristic ultimate tensile strength, fu [which is defined as the minimum value of stress below which not more than a specified percentage of corresponding stresses of samples tested are expected to occur. However, in some countries like the USA, steel is specified according to the (characteristic) yield strength.

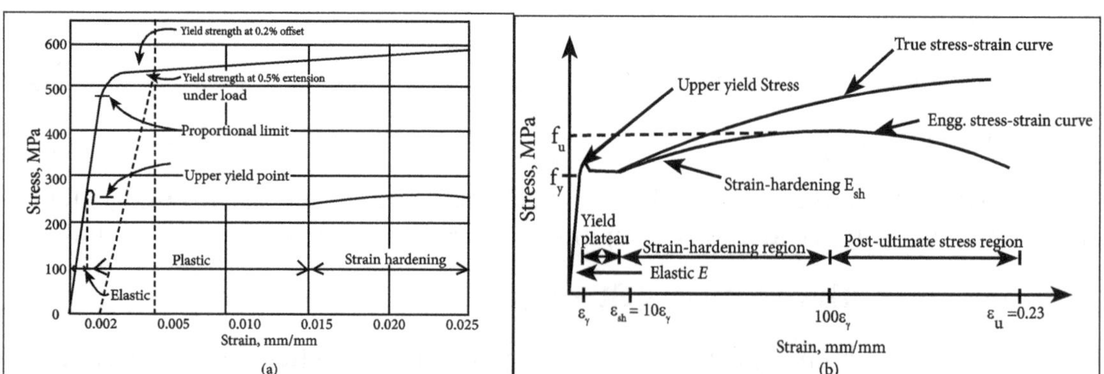

Typical stress-strain curves of mild steel-(a) stress-strain curves of ordinary and high-strength steel, (b) stress-strain curve of a mild steel specimen.

Inelastic Cyclic Response

The stress–strain response of most materials under cyclic loading is different from that under single (monotonic) loading. For fatigue analysis, it is necessary to consider the cyclic material behaviour for strength and life calculations.

When steel is subjected to cyclic loading in the inelastic range, the yield plateau is suppressed and the stress–strain curve exhibits the Bauschinger effect, in which non-linear response develops at a strain much lower than the yield strain, as shown in Figure. As seen from this figure, as the amplitude of response increases, the stress level for a given strain also increases and can substantially exceed the stress indicated by the monotonic stress–strain curve.

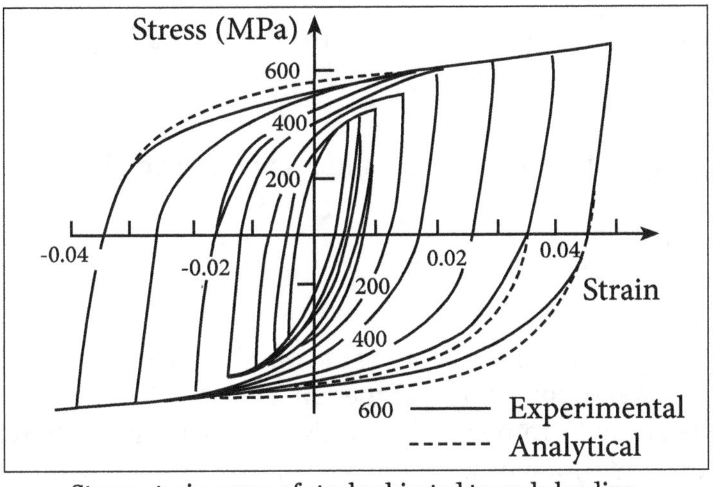

Stress-strain curve of steel subjected to cycle loading.

Characteristic Strength

Variations in material properties should be recognized and taken into consideration in the design process. The material properties that are of greatest importance in the design of structures using steel are yield strength, maximum percentage elongation and Young's modulus. Other properties that are of less importance are hardness, impact resistance and melting point.

This curve can be mathematically expressed by the equation shown in figure which can be used to define 'safe' values for design purposes. When defined, this safe value of yield strength is called characteristic strength. If the characteristic strength is defined as the mean strength then from figure, 50% of the material has a characteristic strength below this value and hence is not acceptable. Hence a characteristic value which has a particular chance (often 95%) of being exceeded in any standard tension test is chosen.

If a number of samples are tested for a particular property (e.g., yield strength) and the number of specimens with the same strength (frequency) are plotted against the strength then the results approximately fit a normal distribution curve as shown in Figure.

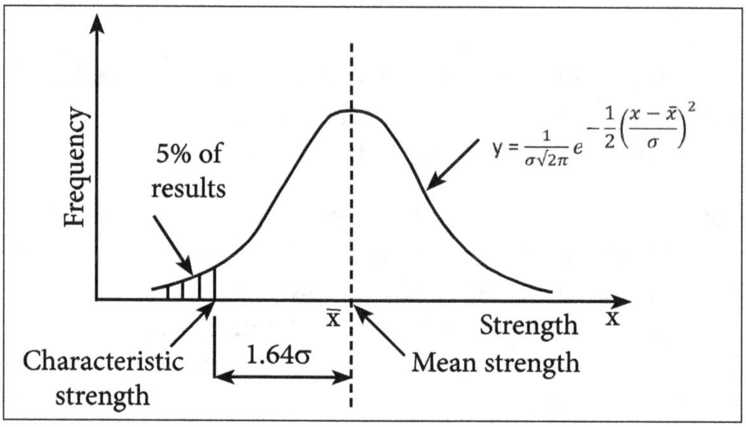

Thus the characteristic strength is calculated from the equation,

$$f_k = f_{mean} - 1.64\,\sigma$$

Where σ is the standard derivation for n samples and is given by,

$$\sigma = \left[\frac{\sum (f_{mean} - f)^2}{(n-1)^{0.5}}\right]$$

The characteristic strength of steel is the value obtained from tests at the rolling mills but by the time the steel becomes part of the finished structure its strength might have been reduced. The strength to be used in design calculations is therefore the characteristic strength divided by a partial safety factor. The value of the partial safety factor adopted for steel is given in Table of IS 800: 2007 as 1.10 for yielding resistance.

Ductility

Ductility is also described as the ability of a material to change its shape without fracture. In other words, the ductility of a structure or its members is the capacity to undergo large inelastic deformations without significant loss of strength or stiffness. The stress–strain curve of a material also indicates the ductility. It is the amount of permanent strain, i.e. strain exceeding proportional limit, up to the point of fracture. The ductility of the tension test specimen is measured by determining the percentage elongation. The specified gauge length according to the code is as follows,

Gauge length $= 5.65\sqrt{A_o}$

$$\text{Percentage elongation } \frac{(\text{elongated length between gauge point} - \text{gauge length}) \times 100}{\text{gauge length}}$$

The measured elongation is influenced by the gauge length, strain rate of test and failure within or outside the gauge length.

Values of 20% can be obtained for mild steel but are less for high-strength steel. By improper testing one may get percentage elongation values of 20–60 and hence the test houses should be careful with the testing process. A high value is advantageous because it allows the redistribution of stresses at ultimate load and the formation of plastic hinges. The minimum required percentage elongations of steel produced in India as per the code are given in Table IS 800: 2007. For most standard mild steels, the values are greater than the minimum required. However, rerolled steel or improperly controlled steel may give higher strength but less percentage of elongation.

Low Temperature and Toughness (Brittle Fracture)

In structural steel design, toughness is a measure of the ability of steel to resist fracture under impact loading, i.e., the capacity to absorb large amounts of energy. Toughness can be an important design criterion, particularly for structures subject to impact loads (e.g., bridges) and for those subject to earthquake loads. Hence both strength and ductility contribute to toughness.

At room temperature, common structural steel is very tough and fails in a ductile manner. At temperatures below 0°C, steel structures sometimes fail suddenly and without warning. A right combination of low temperature, an abrupt change in section size (notch effect) or an imperfection, and the presence of tensile stress can initiate a failure called brittle fracture. This may begin as a crack which may propagate and cause the member to fail. Most brittle fractures occur under static load at stress levels which are not excessive, but they may also be due to the dynamic application of a load or some overload.

One of the best known brittle fractures occurred in Boston in a riveted steel tank of diameter 27.5 m and height 15 m which contained 7.57 million litres of molasses. The tank failed in an explosive manner on January 15, 1919. Similar sudden failures of steel water tanks, oil tanks, transmission lines, ships, plate girders, and bridges have occurred in the past (McGuire 1968). Most of these failures have occurred under normal service conditions, in welded structures, and at low temperatures. During the cold winter of 1977, several spectacular failures occurred in bridge structures in Illinois, Minnesota and Pennsylvania in the USA. On January 22, 1988, several brittle failures occurred in a bridge in Providence, Rhode Island, the USA (Gaylord et al. 1992).

The Charpy V-notch test (CVN test) is commonly used to evaluate the behaviour of metals as they are affected by an abrupt change in cross section (IS 1757). In this test, the temperature of the specimen is varied, the energy absorbed by each specimen is recorded and the energy–temperature curve is plotted. From this curve, a transition temperature corresponding to some level of energy absorption (usually 20 or 27 J) is selected.

Structural steels vary greatly in toughness. Highly killed, fine grain steel with a suitable chemical composition or specially heat-treated steel exhibit considerable toughness.

IS 2026 and IS 1757 codes allow the use of only those steels that exhibit a minimum energy absorption capacity at a predetermined temperature (e.g., 20 J at 23 ± 5°C). In addition to the chemistry of steel, size of plates, residual stress and cold work also affect toughness.

Lamellar Tearing

Lamellar tearing is a form of brittle fracture that may occur in certain welded joints. For example, a tear can occur if a large weld (or welds from both sides) is placed on a thick plate since the shrinkage strains from the welding operation will be large and restrained. The restraint may be developed due to the weld on the far side or due to the member thickness or due to a combination of both the factors.

Generally I-sections are adequately ductile when loaded either parallel or transverse to the rolling direction. A thin stiffened column is susceptible to lamellar tearing since the flange stiffeners that are welded to the column flange produce a restraint. A large overmatch of electrode and base metal in a full penetration butt weld also tends to increase the possibility of tearing.

The use of fillet welds a joint design that allows weld shrinkage to occur in the rolling direction, and the sequence of welding to minimize shrinkage strains are practical methods used to avoid lamellar tearing.

High-Temperature Effects

Steel is not a flammable material. However its strength reduces with rise in temperature. The yield as well as tensile strength at 500°C are about 60–70% of that at room (about 21°C) temperature. The drop in strength is much higher at still higher temperatures.

Hence, steel frames enclosing materials that are flammable require fire protection to control the temperature of steel members for a sufficient time for the occupants to seek safety or for the fire be extinguished before the building collapses. In many cases the building does not collapse even at high temperatures. But the members are deformed beyond acceptable limits, and hence have to be replaced.

The fire-resistance design of steel members, the methods to model real fire and its effect on steel members are discussed in detail in Subramanian (2008).

Hardness

Hardness is a measure of the resistance of the material to indentations and scratching. Several methods are available to determine the hardness of steel and other metals. In all these methods an 'indentor' is forced on to the surface of the specimen. On removal the size of the indentation is measured using a microscope. Based on the size of the indentation, the hardness of the specimen is determined. Brinell hardness (typical value: 150–190) and

Vickers hardness tests are used to determine hardness.

1.1.1 Structural Steel Sections

Various kinds of structural steel sections and their technical specifications are as follows:

- Beams
- Channels
- Angles
- Flats

Steel Beams

Steel Beams is taken into account to be a structural part which mainly carries load in flexure meaning bending. Sometimes beams carry vertical gravitational force however are also capable of carrying horizontal loads usually in the case of an earthquake. The mechanism of carrying load in a beam is extremely unique, like the load carried by a beam is transferred to walls, columns or girders that in turn transfer the force to the adjacent structural compression members. The joists rest on the beam in light-weight frame constructions.

Steel beams.

The beams are known by their profile meaning:

- The length of the beam.
- The shape of the cross section.
- The material used.

The most normally found steel beam is that the I beam or the wide flanged beam also known by the name of universal beam or stouter sections as the universal column. Such beams are normally used in the construction of bridges and steel frame buildings.

The most commonly found types of steel beams are varied and they are mentioned below:

- I beams.
- Wide flange beams.
- HP shape beams.

Beams expertise tensile, shear and compressive stresses internally because of the loads applied to them. Typically below gravity loads there is a slight reduction within the original length of the beam. This leads to a smaller radius enclosure at the top of the beam thus showing compression. While an equivalent beam at the bottom is slightly stretched enclosing a larger radius arc because of tension. The length of the beam midway and at the bends is the same as it is not under tension or compression and is defined as the neutral axis. The beam is entirely exposed to shear stress above the support. There are some reinforced concrete beams that are fully under compression, these beams are called pre-stressed concrete beams and these are built in such a manner to produce a compression more than the expected tension under the loading conditions.

The pre-stressed concrete steel beams have the manufacturing method like, first the high strength steel tendons are stretched and so the beam is cast over them. Then because the concrete begins to cure the tendons are released thus the beam is instantly under eccentric axial loads. An internal moment is created due to the eccentric axial load which in turn will increases the moment carrying capacity of the beam. Such beams are generally used in highway and bridges.

Materials Used

In today's modern construction the beams are generally made up of materials like:

- Steel.
- Wood.
- Reinforced concrete.

Steel Channels

Steel channels are used ideally as supports and guide rails. These are roll-formed products. The most metal used for making channels is steel along with aluminum. There are certain variations that are available in the channels category, the categorization is especially on the shape of the channel, the varieties are mentioned below:

- J channels: This type of channel has two legs and a web. One leg is longer. This channel resembles the letter-J.
- Hat channels: This channel has legs that are folded within the outward direction resembling an old fashioned man's hat.
- U channels: This most typical and basic channel variety. It has a base called as a web and two equal length legs.
- C channels: In this channel the legs are folded back within the channel and resemble the letter-C. C channels are called as rests.
- Hemmed channels: In this type of channel the top of the leg is folded hence forming double thickness.

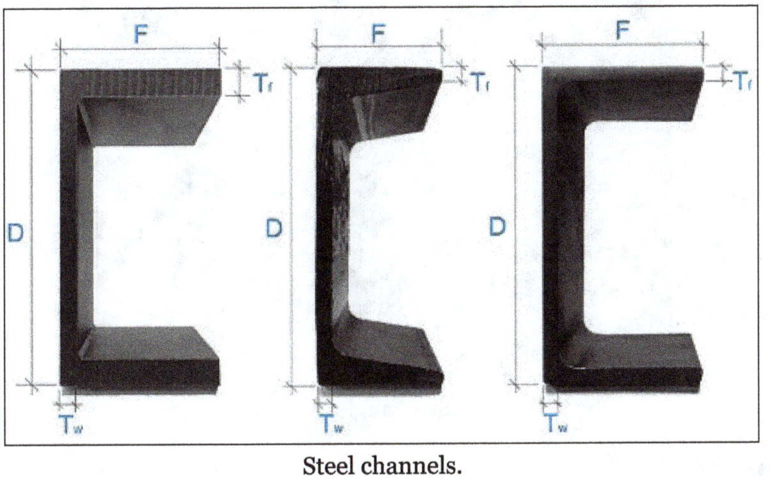

Steel channels.

Application

Steel channels are subjected to a wide array of applications. The application fields are:

- Construction.
- Appliances.
- Transportation.
- Used in making Signposts.
- Used in wood flooring for athletic purposes.

Steel Angle

A steel angle is long steel with reciprocally vertical sides. The steel angles are the most basic kind of roll-formed steel. The most normally found steel angles are formed at a 90 degree angle and has two legs of equal length. The sides are either equal or of various sizes.

There are certain variations within the steel angles depending on its basic construction. The variations are like if one leg is longer than the other then it is called as L angle. If the steel angle is something totally different from 90 degrees then it is called as V angle. In some steel angles, double thickness is achieved by folding the legs inward. If the steel angle has same sides then it means that it has identical width. The steel angles are made according to the strength that is needed for the different structures for construction purposes.

Steel angles.

Applications

The steel angle finds an application in a number of things, they are mentioned below:

- Used in framing.
- Used in trims.
- For reinforcement.
- In brackets.
- Used in transmission towers.
- Bridges.
- Lifting and transporting machinery.
- Reactors.

- Vessels.
- Warehouses.
- Industrial boilers.
- Structural steel angles are used in rolling shutters for fabricating guides for strength and durability.

Steel Flats

Flats are actually thin strips of mild steel having the thickness of the strip normally varying from 12mm to 10mm however thicker flats than this are also available. Steel flats are produced by the use of relatively smooth, cylindrical rolls on rolling mills. Typically the width to thickness ratio of flat rolled products is fairly large. The steel flat bars are manufactured using advanced thickness control technology for controlled thicknesses. The hi-tech machineries enable the production of top grade steel flat bars with superlative flatness and controlled thickness. This product is extremely customized and also the specific sizes according to the client's requirement are produced. After production the flat steels are subjected to a variety of finishes like, painting and galvanizing. The flat carbon steel is a hot or cold rolled strip product also called as a plate product. These plate products have a size variation between 10mm to 200mm and also the thin flat rolled flat rolled product's size varies from 1 mm to 10 mm.

Applications

The steel flats are used in a wide array of applications. The varied applications are listed below:

- Railway parts.
- Ordinance factories.
- Hand tools.
- Engineering industries.
- Auto components-two-wheeler, four-wheeler, commercial vehicles.
- Domestic white goods products.
- Office furniture's.
- Heart pacemakers.
- Tin cans.
- Press working.

1.1.2 Limit State Design Concepts

Steel structures are vital during a variety of land-based applications, including industrial, infrastructural and residential sector. The fundamental strength members in steel structures include support members, plates, stiffened panels/grillages and box girders. Throughout their lifetime, the structures constructed using these members are subjected to various types of loading that is for the most part operational, but may in some cases be extreme or even accidental. Steel-plated structures are probably to be subjected to various types of loads and deformations arising from service requirements that may range from the routine to the extreme or accidental. The mission of the structural designer is to design a structure which will withstand such demands throughout its expected lifetime.

The structural design criteria used for the utility Limit State Design design of steel-plated structures are commonly based on the limits of deflections or vibration for normal use. In reality, excessive deformation of a structure can also be indicative of excessive vibration or noise and so, certain interrelationships can exist among the design criteria being defined and used for separately convenience. The SLS criteria are commonly defined by the operator of a structure or by established practice, the first aim being efficient and economical in-service performance without excessive routine maintenance or down-time. The suitable limits essentially depend on the type, mission and arrangement of structures.

The structural design criteria to stop the Ultimate Limit State Design are based on plastic collapse or ultimate strength. The simplified ULS design of the many kinds of structures has in the past tended to rely on estimates of the buckling strength of components, typically from their elastic buckling strength adjusted by a simple plasticity correction. This is represented by point A in Figure. In such a design scheme based on strength at point A, the structural designer does not use detailed information on the post-buckling behavior of component members and their interactions.

The true ultimate strength represented by point B in Figure is also higher although one will never be sure of this since the actual ultimate strength is not being directly evaluated. In any event, as long as the strength level associated with point B remains unknown, it is difficult to determine the real safety margin. Hence, more recently, the design of structures such as offshore platforms and land-based structures such as steel bridges has tended to be based on the ultimate strength.

The safety margin of structures can be computed by a comparison of ultimate strength with the extreme applied loads (load effects) as depicted in Figure. To obtain a safe and economic structure, the ultimate load-carrying capacity as well as the design load should be assessed accurately. The structural designer may even desire to estimate the ultimate strength not only for the intact structure, but also for structures with existing or premised damage, in order to assess their damage tolerance and survivability.

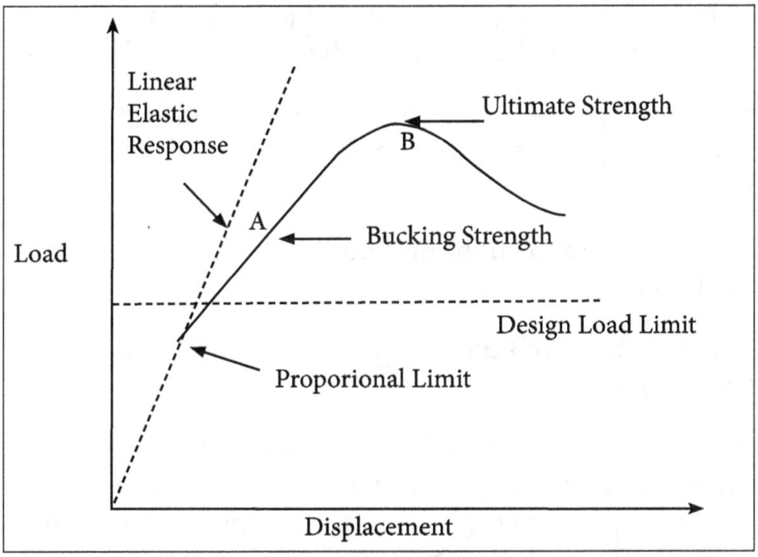

Structural design considerations based on the ultimate limit state.

Ultimate Limit State Design of Steel Structures reviews and describes both fundamentals and practical design procedures in this field. Designs should ensure that the structure does not become unfit/unserviceable for the use for that it is intended to. The state at which the unfitness occurs is known as a limit state.

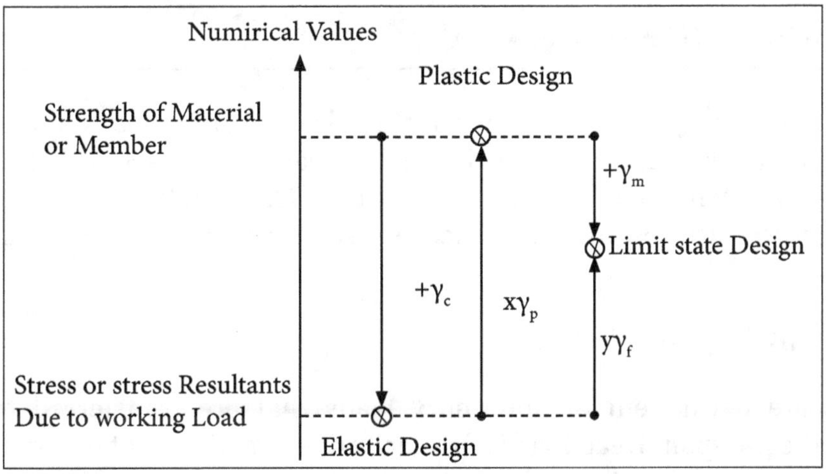

Figure shows the how limit-state design shows separate factors γ_f, that reflects the combination of variability of loading γ_l, material strength γ_m and structural performance γ_p. In the elastic design approach, the design stress is achieved by scaling down the strength of material or member using a factor of safety γ_e as indicated in Figure, while the plastic design compares actual structural member stresses with the effects of factored-up loading by using a load factor of γ_p. Special features of limit state design method are:

- Serviceability and the ultimate limit state design of steel structural systems and their components.

- Due importance has been provided to all probable and possible design conditions that could cause failure or make the structure unfit for its intended use.
- The basis for design is entirely dependent on actual behaviour of materials in structures and the performance of real structures, established by tests and long-term observations.
- The main intention is to adopt probability theory and related statistical methods in the design.
- It is possible to take into account a number of limit states depending upon the particular instance.
- This method is more general in comparison to the working stress method. In this method different safety factors can be applied to different limit states, which is more rational and practical than applying one common factor (load factor) as in the plastic design method.
- This concept of design is appropriate for the design of structures since any development in the knowledge base for the structural behaviour, loading and materials can be readily implemented.

1.2 Loads on Structures

The loads are generally classified as vertical loads, horizontal loads and longitudinal loads. The vertical loads consist of dead load, live load and impact load. The horizontal loads comprises of wind load and earthquake load. The longitudinal loads i.e. friction and braking forces are considered in special case of design of bridges, gantry girders etc.

1. Dead Load

Dead loads are permanent or stationary loads that are transferred to structure throughout the life span. Dead load is primarily due to self-weight of structural members, permanent partition walls, fixed permanent equipments and weight of various materials.

2. Imposed Loads or Live Loads

Live loads are either movable or moving loads without any acceleration or impact. There are assumed to be made by the meant use or occupancy of the building as well as weights of movable partitions or furniture etc. The floor slabs go to be designed to carry either uniformly distributed loads or concentrated loads whichever produce greater stresses in the part under consideration. Since it is unlikely that anyone specific time all floors will not be simultaneously carrying maximum loading, the code permits some reduction in imposed loads in designing columns, load bearing walls, piers supports and foundations.

3. Impact Loads

Impact load is caused by vibration or impact or acceleration. Thus, impact load is equal to imposed load incremented by some percentage referred to as impact factor or impact allowance depending upon the intensity of impact.

4. Wind Loads

Wind load is primarily horizontal load caused by the movement of air relative to earth. Wind load is needed to be thought of in design particularly once the heath of the building exceeds two times the dimensions transverse to the exposed wind surface.

For low rise building say up to four to five storeys, the wind load is not crucial because the moment of resistance provided by the continuity of floor system to column connection and walls provided between columns are sufficient to accommodate the result of these forces. Further in limit state method the factor for design load is reduced to 1.2 (DL + LL + WL) once wind is considered as against the factor of 1.5(DL + LL) when wind is not considered. IS 1893 (part 3) code book is to be used for design purpose.

5. Earthquake Load

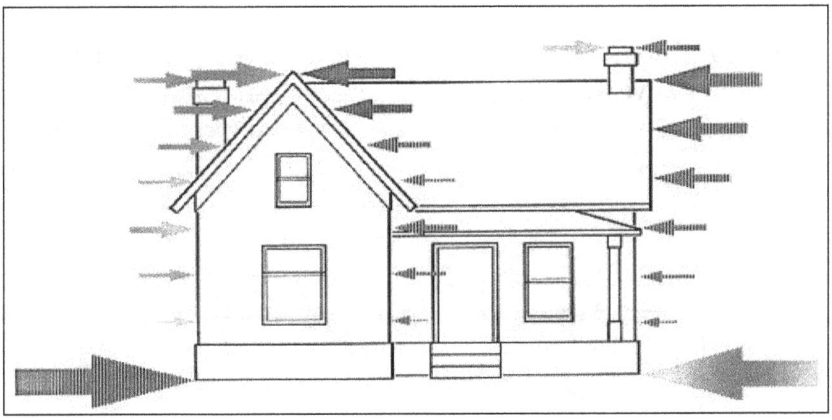

Horizontal earthquake forces (back-and-forth shaking) create 'whipping forces in all parts of a building. These forces must transfer between parts of the building to the foundation.

Earthquake loads are horizontal loads caused by the earthquake and will be computed in accordance with IS 1893. For monolithic reinforced concrete structures set within the seismic zone 2 and 3 while not over 5 storey high and importance factor less than 1, the seismic forces are not crucial.

1.3 Connections using Rivets, Welding and Bolting

Bolted Connection

Bolts can be used in both bearing-type connections and slip-critical connections. Slip-critical connections are used for structures designed for vibratory or dynamic loads such as bridges, industrial buildings and buildings in regions of high seismicity.

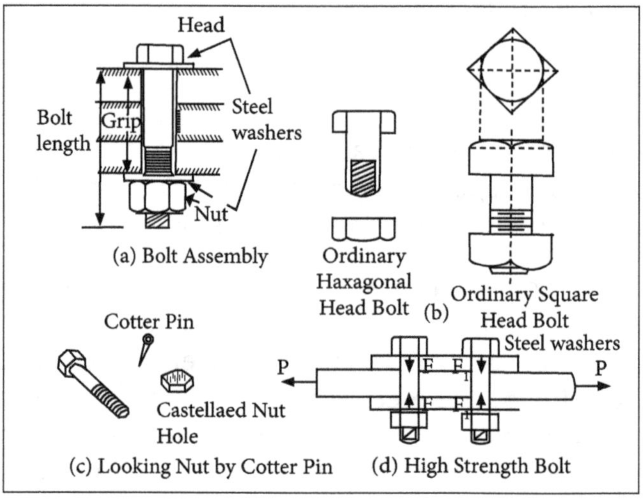

Bolted joints.

Butt Welds

A butt weld is made within the cross-section of the abutting plates in a butt or tee joint. Usually the plate edges have to be prepared before welding, shown in Figure b. In some cases if the plate thickness is less than about 5mm, edge preparation may be avoided, shown in Figure a:

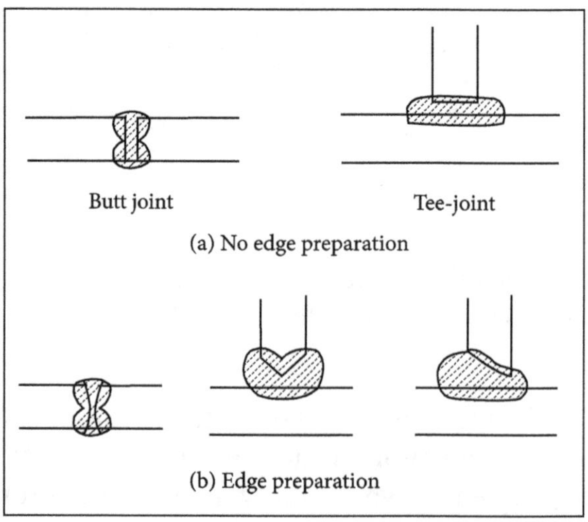

Butt weld with full preparation.

The bevelled plate edges in a butt weld may take various geometrical forms, shown in Figure:

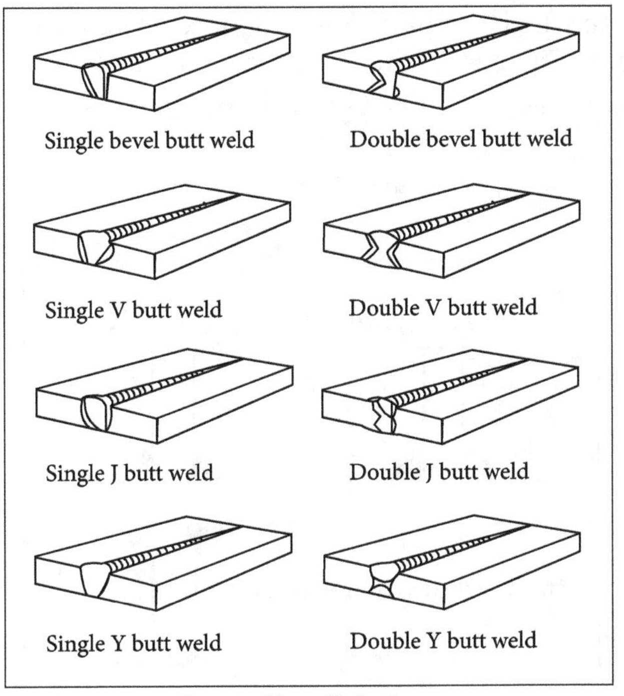

Types of bevelled edges.

For butt welds a distinction is made between:

- Full penetration butt weld in which there is a complete penetration and fusion of weld and parent metal throughout the thickness of the joint, shown in Figure.
- Partial penetration butt weld, in which there is a weld penetration less than the full thickness of the joint, shown in figure:

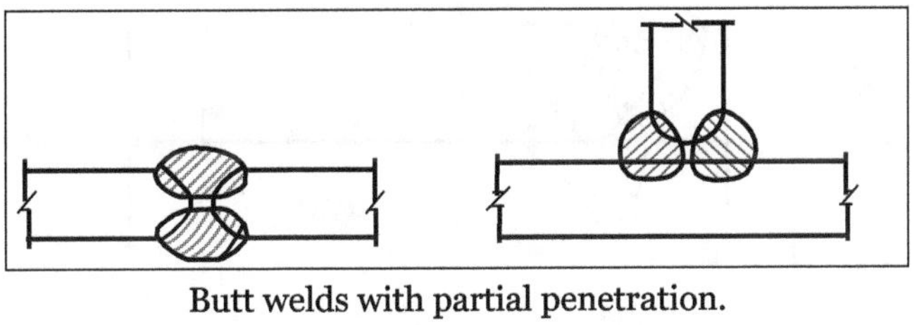

Butt welds with partial penetration.

Fillet Welds

A fillet weld is a weld of approximately triangular cross-section applied to the surface profile of the plates. No edge preparation is required. Therefore fillet welds are usually cheaper than butt welds. According to the relative position of the parts to be welded, there are three types of fillet weld applications:

- Lap joint, in which the parts welded are in parallel planes, shown in Figure a.

- Tee or cruciform joint, in which the parts welded are more or less perpendicular to one another, shown in Figure b.
- Corner joint in which the parts are also more or less perpendicular to one another, shown in Figure c. To improve the strength and stiffness of the joint, the outer corner is normally butt welded, shown in figure:

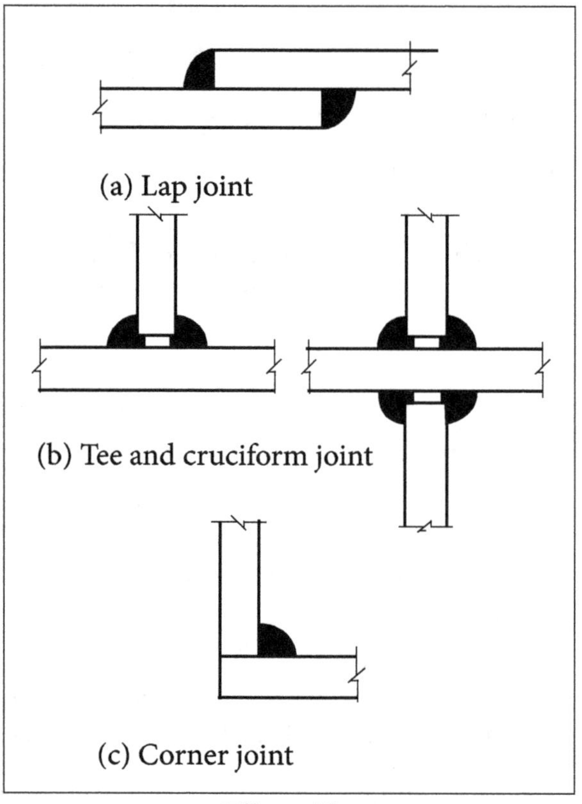

Fillet welds.

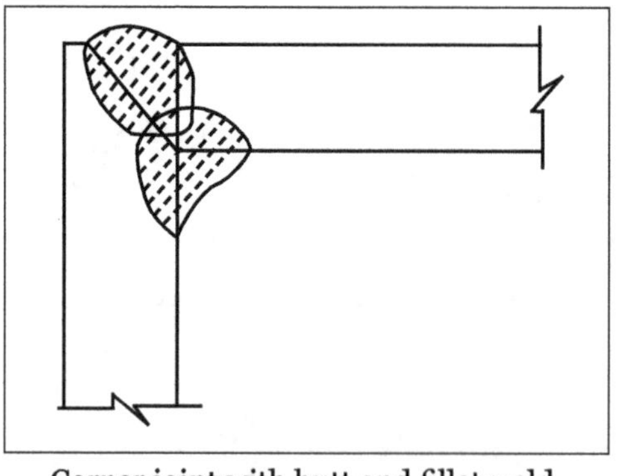

Corner joint with butt and fillet weld.

Fillet welds that can be laid in a single run are particularly economic. In the workshop 8mm welds are often possible but if site welding is to be used this figure might be reduced ex. to 6mm.

Plug and Slot Welds

Slot and plug welds, shown in Figure are seldom used in building structures. They principally prevent buckling or separation of lapped plates.

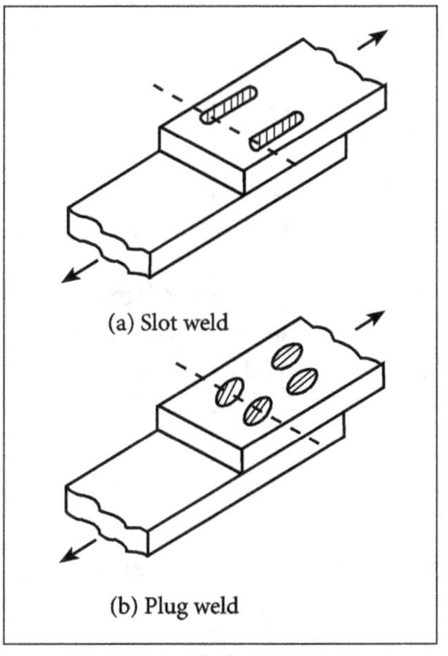

Slot and plug weld.

Spot Welds

Spot welds are seldom used in building structures. The parts to be joined, which must be thin are clamped together by two electrodes shown in Figure. A current passing through the electrodes melts the pieces locally and the clamping pressure forces them together forming a fused spot. A line of such spots constitutes a joint.

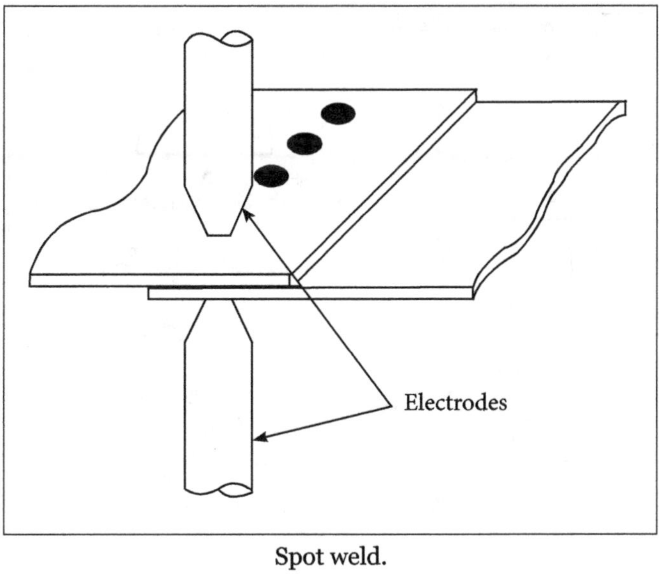

Spot weld.

Efficiency of a Bolted Connection

$$\text{Efficiency}[\eta] = \frac{\text{Strength of Joint}}{\text{Strength of Solid Plate}} \times 100$$

The center to center distance of two adjacent rivets in a row is called pitch.

Staggered Pitch and Prying Action

The distance between any two consecutive rivets in a zigzag riveting, measured parallel to the direction of stress in the member is called staggered pitch.

Moment resisting beam to column connection often contain regions in which the bolts will be require to transfer load by direct tension, such as the upper bolt in end plate connection we should consider an additional force induced in the bolts as a result of so called prying action.

Uses

- Bolts can be used for making end connections in tensions and compression member.
- Bolts can also be used to hold down column bases in position.
- They can be used as separators for purlins and beams in foundations, etc.

Reverting

Reverting is a method of joining together Structural Steel Components by inserting ductile metal pins, called rivets into holes of the components to be connected from departing.

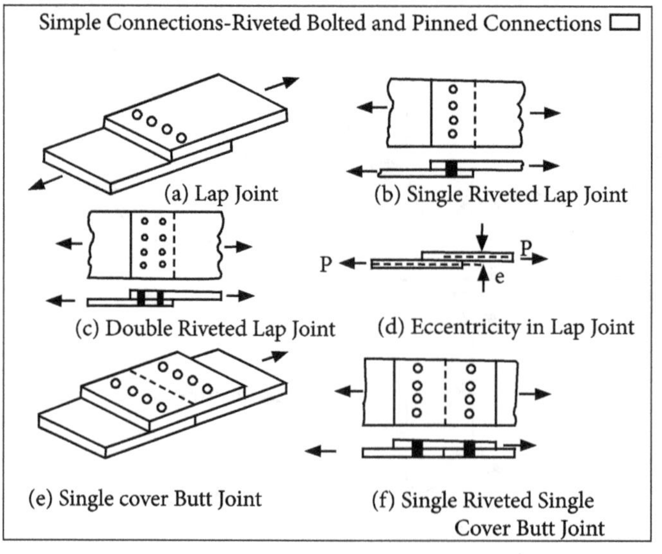

Types of Rivet Joints

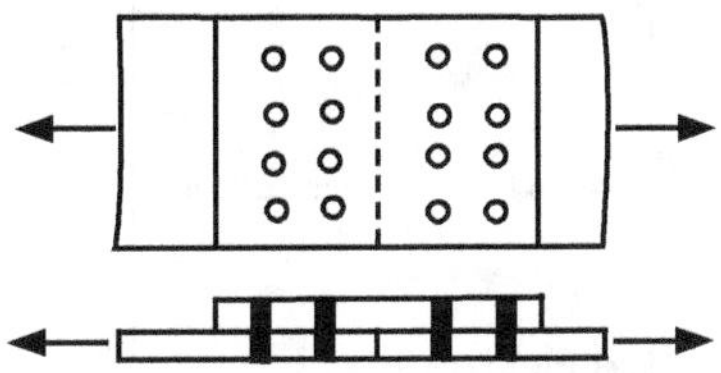

(g) Double Riveted Single Cover Butt Joint

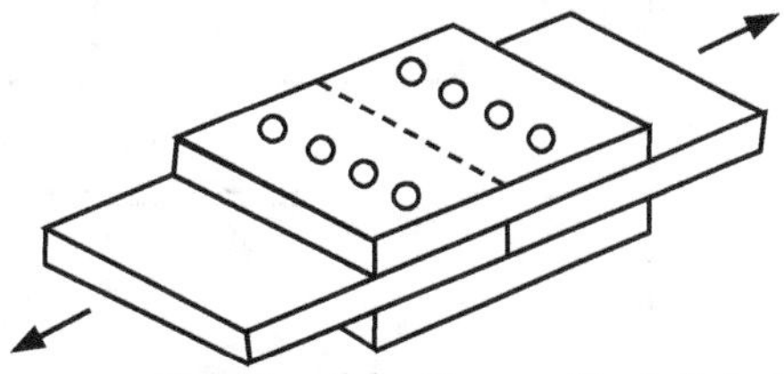

(h) Double Cover Butt Joint

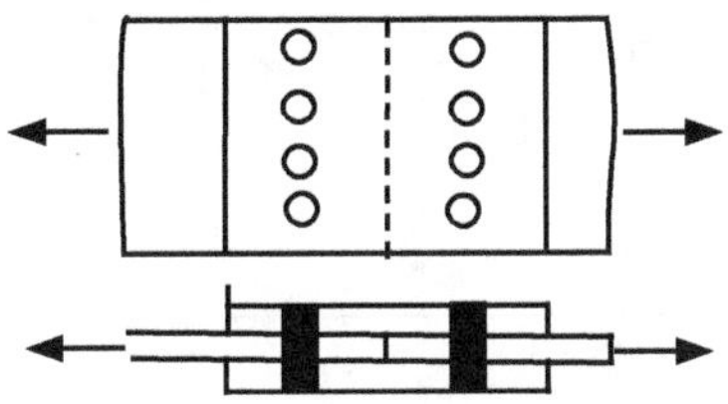

(i) Single Riveted, Double Cover Butt Joint

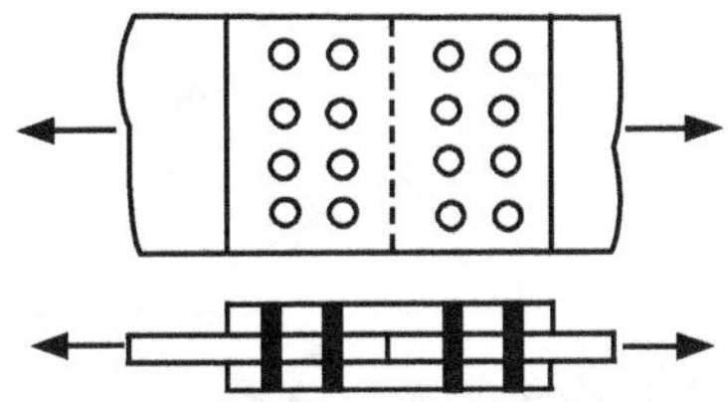

(j) Double Riveted, Double Cover Butt Joint

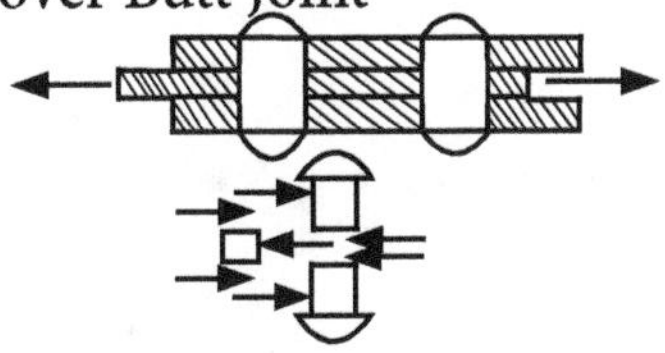

(k) Butt Joint, Rivet in Double Shear

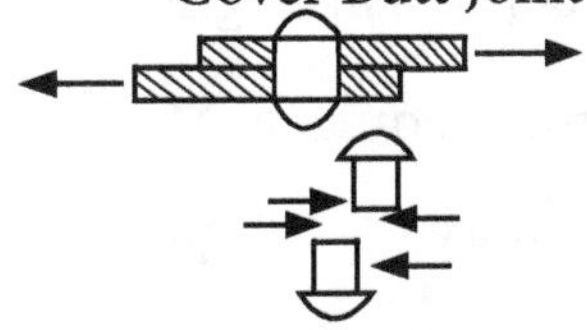

(l) Lap Joint, Rivet in Single Shear

Comparison Between Guage Distance and Pitch of the Bolt

Gauge Distance		Pitch of the Bolt
1.	It is the distance between the two consecutive both of adjacent serves.	It is the centre spacing of the bolts in a row.
2.	It is measured at righit angles to the directi-con of load.	It is measured along the direction of load.

Welded Connection

Structural welding is a process by which the parts that are to be connected are heated and fused, with supplementary molten metal at the joint.

A relatively small depth of material will become molten, and upon cooling, the structural steel and weld metal will act as one continuous part where they are joined.

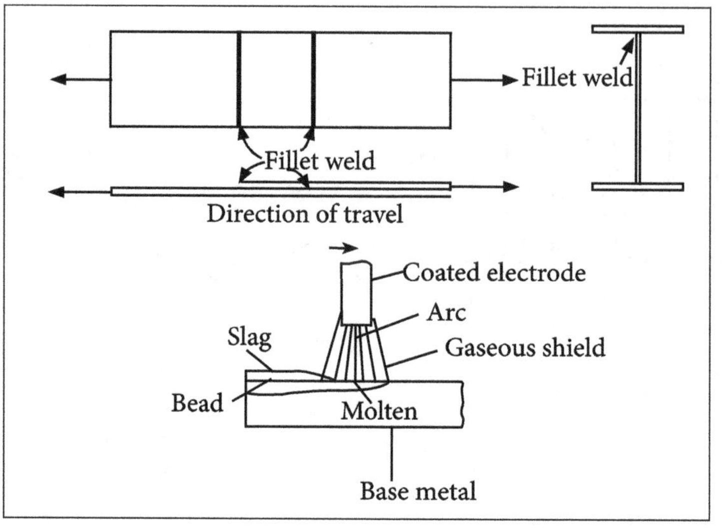

Advantages of Welded Connection

- Welded structures are lighter.
- Welding process quicker.
- Welding is more adaptable.
- It provided good aesthetic appearance.

Disadvantages of Welded Connection

- Brittle fracture is highly possibles.
- It is likely to distort in the process.
- They are over rigid.
- The inspection of welded joints is difficult.

Problems

1. A tie member of a roof truss consists of 2 ISA 10075, 8mm. The angles are connected to either side of a 10 mm gusset plates and the member is subjected to a working pull of 300 kN. Let us design the welded connection and assume connections are made in the workshop.

Solution:

Working Load = 300kN

$\therefore$ Factored Load = 300 × 1.5 = 450 kN.

Thickness of Weld:

(i) At the rounded toe of the angle section, size of weld should not exceed $=\frac{3}{4}\times$ thickness.

(ii) At top (Reference Figure) the thickness should not exceed

$$s = t - 1.5 = 8 - 1.5 = 6.5 \text{ mm}$$

Hence provide s = 6 mm, weld.

Each angle carries a factored pull of $\frac{450}{2} = 225\text{kN}$.

Let L_w be the total length of the weld required.

Assuming normal weld, t = 0.7 × 6 mm

$\therefore$ Design strength of the weld $= L_w t \frac{f_u}{\sqrt{3}} \times \frac{1}{1.25}$

$$= L_w \times 0.7 \times 6 \times \frac{410}{\sqrt{3}} \times \frac{1}{1.25}.$$

Equating it to the factored load,

We get,

$$Lw \times 0.7 \times 6 \times \frac{410}{\sqrt{3}} \times \frac{1}{1.25} = 255 \times 10^3$$

$$\therefore\ L_w = 283 \text{ mm.}$$

Centre of gravity of the section is at a distance 31 mm from top.

Let L_1 be the length of top weld and L_2 be the length of lower weld. To make centre of gravity of weld to coincide with that of angle,

$$L_1 \times 31 = L_2\ (100 - 31)$$

$$\therefore\ L_1 = \frac{69}{31} L_2$$

$$L_1 + L_2 = 283$$

i.e.,

$$L_2 = \left(\frac{69}{31} + 1\right) = 283$$

or,

$$L_2 = 87 \text{ mm.}$$

$\therefore\ L_1 = 195$ mm.

Provide 6 mm weld of L_1 = 195 mm and L_2 = 87 mm as shown in the Figure:

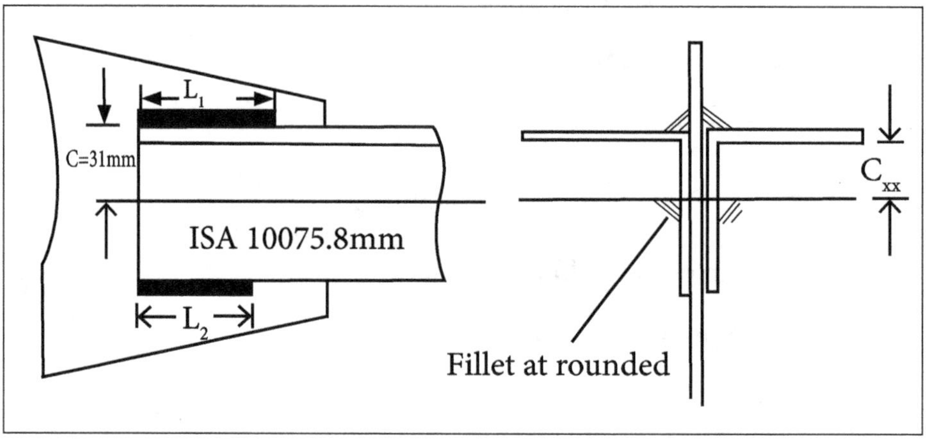

2. ISA90 × 90 × 10mm is connected to a gusset plate 12mm thick by welding. The member carries an axial factored load of 200 kN. Use Fe410 grad steel.

Solution:

Design Weld Connection:

ISA 90 × 90 × 10mm.

Connect to guest plate = 12 mm tk by Welding.

Factored Load = 200 kN.

Weld Connection an all three sides

Use Fe410 grade steel.

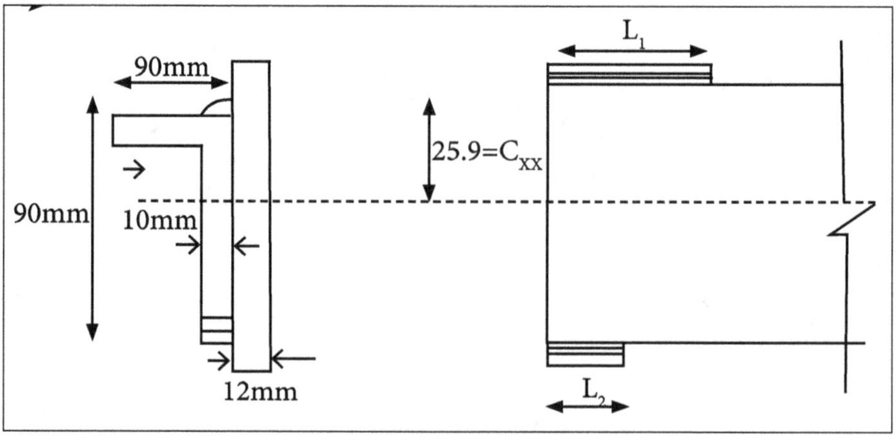

1. Design Load:

Factored Load = 200 kN

2. Size and Throat tk of Weld:

(i) Maximum size of weld @ rounded toe of angle,

$$=\frac{3}{4}[\text{TK of angle}]$$

$$=\frac{3}{4}[10]=7.5\text{mm}$$

(ii) Maximum size of weld = TK of thinner Plate = 1.5m,

$$= 10 - 1.5 = 8.5\text{mm}.$$

(iii) Minimum size = 5 mm; as per Table IS 800 for H of thicker part between 10 to 20 mm.

Let us choose the weld size as 7.5 mm.

Threat tk = 0.7[Size of Weld]

$$= 0.7 \times 7.5 = 5.25\text{mm}.$$

3. Strength of Weld:

Pull transmitted by 1mm length of weld,

$$= A_g f_{wd} \text{ [as Per Cl. 10.5.7.1.1 of IS800:2007]}$$

$$f_{wd} = \frac{f_{wn}}{\gamma_{mw}} = \frac{\left(f_u/\sqrt{3}\right)}{\gamma_{mw}}$$

$$= \frac{\left[410/\sqrt{3}\right]}{1.25} \Rightarrow 189.38\text{MPa}.$$

Strength/mm of weld,

$$= \frac{1[0.7\times 7.5]\left[410/\sqrt{3}\right]}{1.25}$$

$$= 994\text{N}.$$

4. Length on Weld:

L_w = Total Length of weld on top and bottom on each angle.

Force transmitted by each Angle.

Factored Load = Design Strength of Weld $= 200 = 200$ kN,

$100 \times 3 = 994\, L_w$

$L_w = 201.6 \approx 201$ mm.

5. Length on each Side of One Angle:

Centre of Gravity of ISA 90 × 90 × 10mm lies @ 25.9 mm from top.

Let L_1 be the Length of weld @ top and L_2 be the Length of weld@bottom.

Forces above C.G of angle = Forces below C.G of angle,

$$L_1[25.9] = L_2[90 - 25.9]$$

$$L_1 = \frac{90-25.9}{25.9} L_2$$

$$L_1 = 2.47\, L_2$$

$$L_\omega = L_1 + L_2$$

$$201 = 2.47\, L_2 + L_2$$

$$201 = 3.47\, L_2$$

$$L_2 = \frac{201}{3.47} = 57.9 \text{ mm.}$$

$$L_1 = 201 - 57.9$$

$$L_1 = 143.1 \text{ mm.}$$

∴ Provide 7.5 mm weld of Length, L_1 = 143.1 mm @ top and Length, L_2 = 57.9 mm @ bottom on each of the angle.

3. Let us discuss the problem based on riveted joint. A plate in which the axial load tension is 520 kN to be provided with a splice joint. The size of the plate is 400 mm × 20 mm.

Solution:

Size of plate = 400 mm × 20 mm

Force P = 520 kN

Permissible shear stress = 80 N/mm^2

Permissible bearing stress = 250 N/mm²

Permissible tensile strength of plate = 150 N/mm².

Shear strength of one 20 mm ø rivet $= 2\frac{\pi d^2}{4}$ is,

$$= \frac{2(\pi)(21.5)^2}{4}(20)$$

$$= 58058 \text{ N} \quad \dots (1)$$

Permissible bearing strength of one rivet,

$$= dtf_b$$
$$= 21.5\ (20)\ (250)$$
$$= 107500 \text{ N} \quad \dots (2)$$

∴ Strength of one rivet = Least value of shearing and bearing strength of one river,

$$= 58058 \text{ N}$$

∴ No. of rivets requires $= \frac{520\ (1000)}{58058}$

$$= 8.95 \approx 9 \text{ rivets.}$$

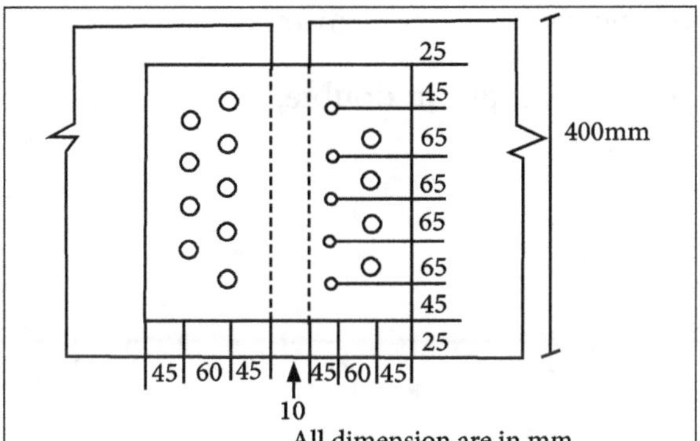

All dimension are in mm

Thickness of cover plate required.

Let the width of cover plate,

= 400 mm

= 400 – 50 – 2(21.5)

= 242.5 mm.

Net area of cross section of each cover plate,

$$=\frac{P}{2(f_t)}$$

$$=\frac{520000}{2(150)}=1733.33mm^2$$

Thickness of cover plate,

$$=\frac{1733.33}{242.5}$$

$$=7.14 \text{ mm}$$

$$=45+60+45+10+45+60+45$$

$$= 310 \text{ mm.}$$

∴ Provide a double rivet double cover joint rivet plate size 310 × 242.5 × 8 mm with rivet 2 rows.

4. The double riveted double cover butt joint to connect 2 plate of 12 mm thick. Adopt power driven rivets take f_y = 250 MPa. Let us determine also the efficiency of the joint.

Solution:

- Shearing stress of one rivet = 100N/mm2.
- Bearing stress of one rivet = 300 N/mm2.

∴ Shear strength of rivet bitch length in double,

$$=2\left[2\left(\frac{\pi d^2}{4}\right)f_s\right]$$

Rivet Cover Joint,

$$=2\left[2\left(\frac{3.4\times(23.5)^2}{4}\right)100\right]$$

$$=173407 \text{ N}$$

$$=2\ (d_z\ f_b)$$

$$=2\ (23.5\ (12)\ 300$$

$$=169200 \text{ N.}$$

Rivet value = 169200 N.

Fearing Strength Pitch length = (p – d) t f_t,

$$=(p - 23.5)\ (12)\ (250)$$

$$=(p - 23.5)\ (12)\ (250)=169200$$

$$=P=56.4 + 23.5$$

$$=79.9 \text{ say } 80 \text{ mm}$$

Strength of solid plate = Pt f_t,

$$=f_o\ (12)\ (250)$$
$$=240000 \text{ N}$$

$$\therefore \text{ Efficiency of solid joint } = \frac{169200}{240000} \times 100$$

$$= 70.5\%.$$

5. Here we solve a single riveted double cover butt joint is used to connect two plate 12 mm thick. The rivets used as power driven 18 mm in diameter at pitch of 60 mm.

Solution:

- Since it is power driven rivets, let us assume the following permissible stress values.
- Axial tensile stress = 100 N/mm2.
- Shear stress = 100 N/mm2.
- Bearing stress = 300 N/mm2.

Since the joint is double cover butt joint the rivets are in double shear. Since, the joint is a single riveted joint. The number of rivets covered per pitch length = 1 finish diameter of the rivert hole = 18 + 1.5 = 19.5 mm.

Considering on pitch length of the joint:

(i) Tearing strength per pitch length,

$$P_t = (p-d)\ t\ ft$$
$$=(60-19.5)\ (12)\ (100)$$
$$=48600\text{N}.$$

(ii) Shearing strength per pitch length,

$$=(1)(2)\left(\frac{\pi d^2}{4}\cdot fs\right)$$

$$=(1)(2)\left[\frac{\pi(19.5)^2}{4}(100)\right]$$

$$=59699 \text{ N.}$$

(iii) Bearing strength per pitch length,

$$=(1)\ dt\ fb$$

$$=(1)\ (19.5)\ (12)\ (300)$$

$$=70200 \text{ N.}$$

∴ Safe load per pitch length = 48600 N.

Strength of solid plate per pitch length = p_t ft,

$$= 60(12)\,(100)$$

$$= 72000 \text{ N.}$$

$$\therefore \text{ Efficiency of the joint } = \frac{48600}{72000}\times 100$$

$$= 67.5\%.$$

1.3.1 Design of Bolted and Welded Joints

High Tension Bolts

High tension bolts are developing a high initial tension which facilitates clamping or gripping the components converted.

Spitting of Plates

The failure mode of two connected plate due to tearing or shearing is called spitting of plate.

Type of Bolts

Some type of bolts are:

- Unfinished bolts or black bolts or 'C' grade bolts.

Turned bolts such as:

- Precision bolts or 'A' grade bolts.
- Semi-precision or 'B' grade bolts.
- Rubbed bolts or fluted bolts.
- High strength friction grip bolts.

Problems

1. Let us discuss the doubly bolded lap joint for plates 16 mm thick to carry its full load. Take permissible axial tension in plate 150 N/mm².

Solution:

Design a Doubly Bolded Lap Joint:

Plate = 16 mm thick

Permissible axial tensile, f_u = 150 N/mm².

1. Bolt Diameter and Area of Bolt at Root:

Let us assume nominal dia. 20 mm are used for connections.

Area of bolt @ root,

$$A_{nh} = 0.78\, A_{sb}$$
$$= 0.78 \frac{\pi}{4} d^2$$
$$= 0.78 \frac{\pi}{4} 20^2$$

Answer = 245 mm².

2. Design Strength in Shear:

$$V_{dsb} = \frac{V_{sb}}{\gamma_{mb}}$$

$$V_{nsb} = \frac{f_{ub}}{\sqrt{3}} [n_n A_{nb} + n_S A_{sb}]$$

$$V_{dsb} = \frac{400}{\sqrt{3}} [(1 \times 245) + (0 \times 314)]$$
$$= 56580 \text{ N}$$

γ_{mb} = 1.25 as per Table of IS 800

f_{ub} = 400 MPa

Grade 4.6 bolts,

$$n_s = 0$$

$$n_n = 1$$

$$V_{dsb} = \frac{56580}{1.25}$$

$$= 45266 \text{ N.}$$

3. Design Strength in Bearing [Vdpb]:

$$V_{dsb} = \frac{V_{npb}}{\gamma_{nub}}$$

$$= \frac{2.5\ k_b\ dt\ \delta_u}{\gamma_{mb}}$$

Where k_b is the least of the following,

$$\frac{e}{3d_o} = \frac{1.5\left[20\right]}{3 \times 22} = 0.4545$$

$$\frac{P}{3d_o} - 0.25 = \frac{2.5 \times 20}{3 \times 22} - 0.25 = 0.508$$

$$\frac{f_{ub}}{f_u} = \frac{400}{410} = 0.975$$

1.0

$$K_b = 0.4545$$

$$V_{npb} = 2.5 \times 0.4545 \times 20 \times 16 \times 410$$

$$V_{dpb} = \frac{149076}{1.25}$$

$$V_{dpb} = 119260 \text{ N.}$$

Bolt Value is least of above two.

Bolt Value = 45266N.

4. Full Strength of Solid Plate Governed by Yielding:

Strength of Solid plate per Pitch Length,

$$T_{dg} = \frac{A_g f_y}{\gamma_{m_o}} \text{ \{as per C1.6.2 of IS800-2007\}}$$

$$A_g = P_t$$

$$= 50 \times 16$$

$$A_g = 800 \text{ mm}^2$$

$$f_y = 250 \text{ MPa.}$$

$$f_{M_o} = 1.1[\text{as per Table of IS } 800-2007]$$

$$T_{dg} = \frac{[800] \times 250}{1.1} = 181818.2 \text{ N.}$$

$$\text{No. of bolts required} = \frac{\text{Strength of Solid Plate}}{\text{Bolt Value}}$$

$$= \frac{181818.2}{45266} \approx 4 \text{nos}$$

$$\text{No. of bolts} = 4 \text{ Nos.}$$

2. Let us discuss the problem based on lap joint. A 100 mm × 10 mm plate is to be welded to another plate 150 mm × 10 mm by the fillet welding on three sides. The size of the weld is 6 mm. Take allowable tensile stress in plate equal to 150 N/mm² and allowable stress in weld as 110 N/mm².

Solution:

Design Load,

$$\text{Strength of Plate} = \frac{A_g f_y}{\gamma_{m_o}} = \frac{[100 \times 10] \times 250}{1.1}.$$

As for Table of IS800-2007

Strength of Plate = 227272.7 N.

Size and Threat Thickness of Weld:

Size of Weld = 6 mm.

Throat Thickness,

= 0.7 × Size of Weld.

$= 0.7 \times 6 = 4.2$mm.

Strength of Weld Per 1 mm:

Strength of Weld/mm Length = A_g [f_{wd}]

As per IS800-2007 – CI.10.5.7.1.1,

$$f_{wd} = \frac{\left[f_u/\sqrt{3}\right]}{\gamma_{m_w}}$$

$$= \frac{410}{1.25 \times \sqrt{3}} = 189.2820 \text{ N/mm}^2$$

Strength of Weld /mm = 1 × 4.2 × 189.282

= 795.98 N.

Length of Weld Required,

Lw × 794.98 = 227272.7

Lw = 286 mm.

But length for fillet weld available on sides for 100 mm plate,

= 2 × [100 – 2 × 6]

= 176 mm.

∴ Force transmitted by 176 mm length of fillet,

Weld = 176 × 285

= 50336 N.

The remaining strength is to be transmitted through slot welds.

∴ Strength to be transmitted through Slot Welds = 227272.7

= 176936.7 N.

Area of Slot Weld Required:

$$\text{Area of Slot Weld} \times \frac{250/\sqrt{3}}{\gamma_{m_w}} = 176936.7$$

Area of Slot Weld = 1532mm^2.

Hence let us provide two slot welds of size 50 × 20 mm each giving an area.

= 2[50 × 20] = 2000 1526 mm^2.

3. Here we will discuss about a tie member consisting of angle section 80 mm × 50 mm × 8 mm is welded to a 8 mm gusset plate.

Solution:

Step 1: Determination of full strength of the member ISA 80 mm × 50 mm × 8 mm. From steel table c/s area of ISA 80 × 50 × 8 mm = 978 mm^2.

Assume yield strength of 250 MPa for the member partial safety factor for angle section as per Table of IS 800-2007 is,

$$\gamma_{mo} = 1.1.$$

∴ Design Strength of the member,

$$= \frac{A_g fy}{\gamma_{mo}}$$

$$= \frac{978(250)}{1.1}$$

$$= 222273 \text{ N}$$

Step 2: Size and throat thickness of Weld.

(i) Maximum size of weld at rounded toe of angle,

$$= \frac{3}{4}(\text{Thickness of Angle})$$

$$= \frac{3}{4} = 6 \text{ mm}$$

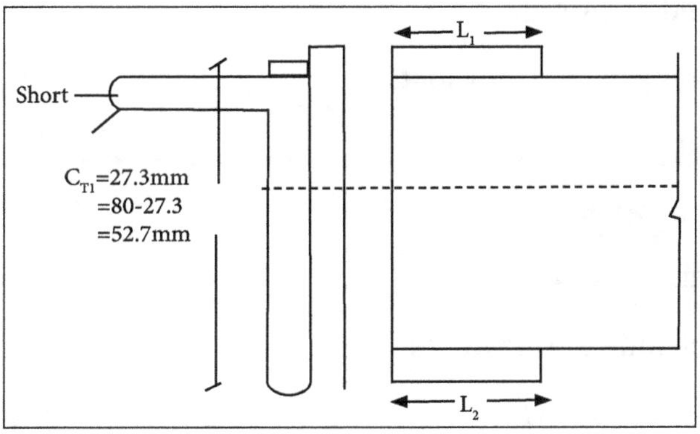

(ii) Maximum Size of Weld,

= Thickness of thinner plate – 1.5 mm

= 8 – 1.5 = 6.5 mm.

(iii) Minimum size of weld = 3 mm.

(as per Table of IS 800-2007 for thickness of thicker up to and including 10 mm).

∴ let the Weld size as 6 mm.

Throat thickness,

$$= 0.7 \text{ (size of weld)}$$

$$= 0.7\,(6) = 4.2 \text{ mm.}$$

Step 3: Strength of weld per 1 mm length.

$$f_{wd}\,\frac{f_{wn}}{\gamma_{mw}}\,\frac{\left(fu/\sqrt{3}\right)}{\gamma_{mo}}.$$

Assume:

$$f_u = 250 \text{ MPa}$$

$$\gamma_{mw} = 1.25$$

$$f_{wd} = \frac{\left(250/\sqrt{3}\right)}{1.25} = 115.47 \text{ MPa.}$$

∴ Strength/mm of Weld = 1(0.7) (6) (115.47)

$$= 484.99$$

$$\approx 485 \text{ N}$$

Step 4: Length of Weld required.

Equate the design strength to factor load.

We get,

$$L_w(\,485) = 222273 \text{ N}$$

$$L_w = 458.31 \text{ mm} \approx 459 \text{ mm.}$$

Step 5: Length of each side of one angle.

$$L_1\,(27.3) = L_2\,(80 - 27.3)$$

$$L_1 = 1.9342$$

Also,

$$L_1 + L_2 = L_w = 459 \text{ mm}$$

$$1.93\ L_2 + L_2 = L_2\ (1.93 + 1) = 459$$

$$L_2 = \frac{459}{2.93} = 237.77 \approx 238\ \text{mm}$$

$$L_1 = 1.93\ (238) = 459.34 \approx 460\ \text{mm.}$$

∴ Provide 6 mm weld of length L_1 = 460 mm at top and length L_2 = 238 mm at bottom.

4. An I.S.L.C. 300 @ 331 N/m is used to transmit a force of 600 kN. The channel section is connected to a gusset plate 10 mm thick overlap is limited to 350 mm. Use slot welds if required.

Solution:

Relevant properties of I.S.L.C. 300 @ 331 N/m are,

$A = 4211\ \text{mm}^2$

$T = 11.6$ mm

$t = 6.7$ mm

As slot welds are required, the maximum size of weld,

= 6.7 – 1, 5

= 5.2 mm

Provide 5 mm-size weld.

Throat thickness = 0.7 × 5 = 3.5 mm

Strength of weld/mm length = 3.5 × 108

= 378N

$$\text{Length of weld required} = \frac{600 \times 1000}{378}$$

= 1587.30

= 1590 mm

The maximum length of the weld that can be provided is,

300 + 2 × 350 = 1000 mm (1590 mm)

Hence, the length of the weld will be provided with slot welds.

Provide width of slots 25 mm (3t = 3 × 6.7 = 20.1 mm or 25 mm whichever is greater).

Let length of the slot be x, $1590 = 2 \times 350 + 300 + 4x$

$$x = 147.5 \text{ mm} = 150 \text{ mm}$$

Provide 150mm long fillet welds in slots as shown in figure.

Example:

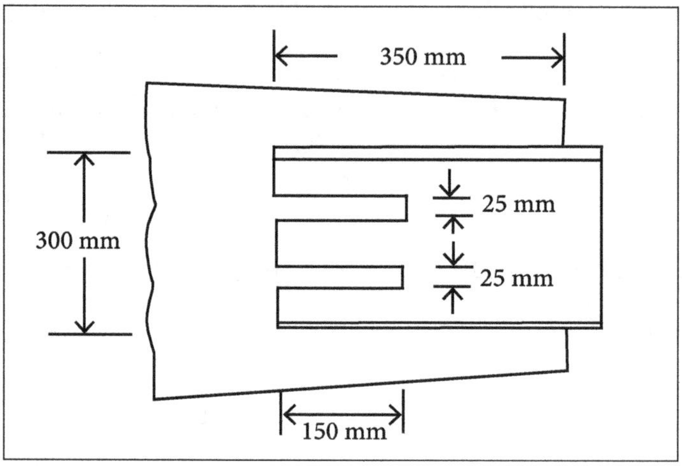

5. The main plates are 12 mm thick connected by 18 mm diameter rivets at a pitch of 100 mm. Let us design the cover plates and also find the percentage reduction in the efficiency of the point if the plates are lap jointed.

Solution:

Since the joint is a butt joint with two cover plates, the rivets are in double shear. Since the joints is single riveted joint, the member of rivets covered per pitch length = 1. Finished diameter cf rivet, d = 22 + 1.5 = 23.5 mm.

Consider one pitch length of the joint.

Tearing strength per pitch length,

$$= P_t = (P - d)t/t$$

$$= (100 - 23.5)\ 16 \times 150 = 1,83,600 \text{ N.}$$

Shearing strength Per Pitch Length:

$$= [1] \times 2fs\ \frac{\pi d^2}{f}$$

$$= 1 \times 2 \times 100 \times \frac{(2.3)^2}{4} = 86,740 \text{ N.}$$

Bearing Strength Per Pitch Length:

$= [1] \times f_b\, dt = 1 \times 300 \times 23.5 \times 16$

$= 1, 12, 800$ N.

Safe load per. pitch length = 86,740 N

Strength of the solid plate per pitch length

$= p_t f_t = 100 \times 16 \times 150 = 2, 40, 000$ N

$$\text{Efficiency of the joint} = \frac{86{,}740}{2{,}40{,}000} \times 100$$

$$= 36.14\%.$$

Result:

Safe load per pitch length = 86,740 N

Efficiency of the joint = 36.14%

1.3.2 Eccentric Connections

If the load is not passing through the CG of the connection. Eg. Bracket connection and seat connection.

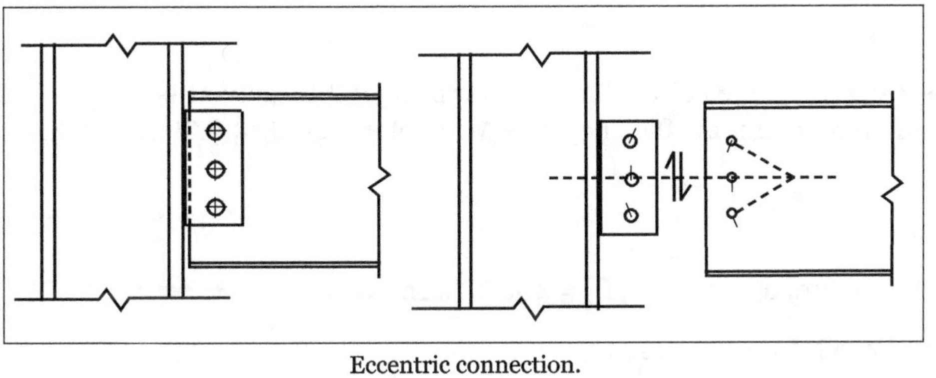

Eccentric connection.

1.3.3 Efficiency of Joints

A numerical value expressed as the ratio of the strength of a riveted, welded or brazed joint to the strength of the parent metal.

Efficiency of Riveted Joints

Efficiency of the single riveted joint can be obtained as ratio between the maximum of P_1, P_2 and P_3 and the load carried by a solid plate which is $s_t pt$.

Thus,

$$\text{efficiency}\ (\eta) = \frac{\min\{P_1, P_2, P_3\}}{s_t pt}$$

In a double or triple riveted joint the failure mechanisms may be more than those discussed above. The failure of plate along the outer row may occur in the same way as above. However in addition the inner rows may fail. For example in a double riveted joint the plate may fail along the second row. But in order to do that the rivets in the first row must fail either by shear or by crushing. Thus the maximum allowable load such that the plate does not tear in the second row is,

$$P_4 = s_t (p-d) t + \min\{P_2, P_3\}.$$

Further the joint may fail by:

- Shearing of rivets in both rows.
- Crushing of rivets in both rows.
- Shearing of rivet in one row and crushing in the other row.

The efficiency should be calculated taking all possible failure mechanism into consideration.

Problems

1. Let us discuss a single bolted double cover butt joint to convert boiler plates of thickness 12 mm for maximum efficiency. Use M16 bolts of grade 4.6. Boiler plates are of the 416 grade.

Solution:

$d = 16$ mm, $d_o = 18$ mm, $f_{ub} = 400\ N/mm^2$

$f_u = 410\ N/mm^2$, $t = 12$ mm.

Since it is double cover but joint, the both are in double shear are section at shank and another at root.

Nominal strength of a bolt in shear,

$$= \frac{400}{\sqrt{3}} \left(\left(1 \times \frac{\pi}{4} \times 16^2 \right) + \left(1 \times 0.78 \times \frac{\pi}{4} \times 16^2 \right) \right)$$

$$= 82651\ N$$

Design strength in shear $= \frac{82651}{1.25} = 66121$ N ...(1)

Provide cover plates of 8 mm thickness.

Design strength of plate per pitch width,

$$= \frac{0.9 \times 410\,(P-18) \times 12}{1.25}$$

$$= 3542.4\,[P - 18] \quad ...(2)$$

Equating (1) and (2) to get maximum efficiency,

$$= 3542.5\,(P - 18) = 66121$$

$$P = 86.67$$

2. A lap joint to connect two plates of thickness 10mm and 12mm respectively to carry a factored load of 150 kN. Use 16mm diameter 4.6 grade bolts and Fe_{410} grade steel.

Solution:

Design a Lab Joint:

- Connect ⇒ two plates of tk = 10 mm and 12 mm.
- Factored Load, P = 150kN.
- dia of bar = 16mm.
- Grade of bolt = 4.6.
- Fe410 grade Steel.
- Efficiency =?

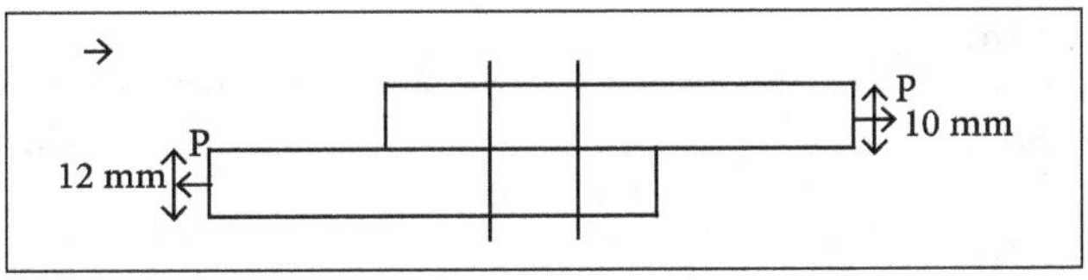

- Nominal diameter of the bolt, d = 16 mm.
- Diameter of the hole, do = 16 + 2 = 18mm.

1. Bolt Diameter and Area of Bolt at Root [Anb]:

$$A_{nb} = 0.78\left[\frac{\pi[16]^2}{4}\right] = 157\ \text{mm}^2.$$

2. Strength of 1 Bolt in Single Shear, Vdsb

$$V_{dsb} = \frac{V_{nsb}}{\gamma_{mb}}$$

$$V_{nsb} = \frac{\delta_{ub}}{\sqrt{3}}\left[n_n A_{nb} + n_s\ A_{sb}\right]$$

As per IS Ran – 2007; Cl 103.3.

fub = 400 MPa [for M16 and 4.6 grade bolts as per IS800.2007 –Table].

$n_n = 1$; $A_{nt} = 157\text{mm}^2$, $n_s = 0$.

$$V_{nsb} = \frac{400}{\sqrt{3}}\left[1(157)+0\right] = 36259\ \text{N}$$

$$V_{dsb} = \frac{36259}{1.25} = 29007\ \text{N}.$$

3. Design Strength of 1 bolt in Bearing [Vdpb]:

$$V_{dpb} = \frac{V_{npb}}{\gamma_{mb}}$$

$$V_{npb} = 2.5\, k_b\ d\, t\, f_u$$

kb is smaller of $\frac{e}{3d_o}, \frac{P}{3d_o} - 0.25, \frac{S_{ub}}{f_u}$ and 1

Minimum end distance e = 1.5d

= 1.5 × 16

[as per IS 800-2007, Cl -10.2.4.2]

e = 2400 ≈ 30 mm.

Minimum Pitch = 2.5 d

= 2d [16] ≈ 40 mm.

$$k_b = \frac{30}{3[18]}, \frac{40}{3[18]} - 0.25; \frac{400}{410}$$

$$= 0.56,\ 0.491,\ 0.976.$$

$$k_6 = 0.491,\ d = 16\ \text{mm},\ t = 12\ \text{mm},\ f_u = 410\ \text{MPa}$$

$$V_{npb} = 2.5 \times 0.491 \times 16 \times 10 \times 410$$

$$V_{npb} = 80524\ \text{N}.$$

$$V_{dpb} = \frac{V_{npb}}{\gamma_{mb}} = \frac{80524}{1.25} = 64419.2 \text{ N.}$$

Bolt Value = Least of 29007 N and 64419.2 N

Bolt Value = 29007 N.

4. Number of Bolts Required:

$$N = \frac{150000}{29007} = 5.17 \approx 6 \text{ bolts.}$$

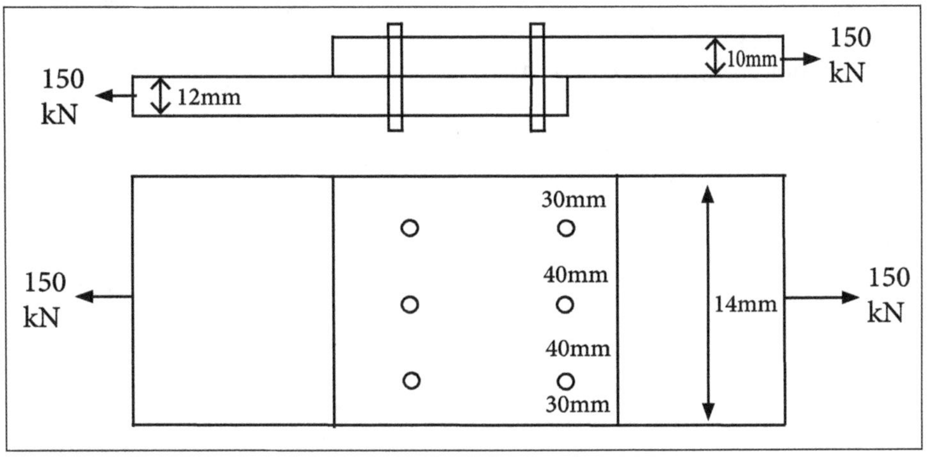

5. Efficiency:

$$\eta = \frac{\text{Strength of Joint}}{\text{Strength of Solid Plate}} \times 100$$

$$= \frac{29007}{\left[\frac{f_y A_g}{200}\right]} \times 100 = \frac{29007 \times 6}{\left[\frac{250 \times 40 \times 10}{100}\right]} \times 100 = 54.69\%.$$

Minimum Pitch = 2.5 × 16 = 40 mm.

Provide bolts at P = 40 mm.

Check for Strength of Bolt in Bearing:

Kb is the minimum of $\frac{e}{3do}, \frac{P}{3do} - 0.25, \frac{f_{ub}}{f_u}, 1.0.$

Assuming sufficient 'e' will be provided,

$K_{bo} = 0.4907.$

Design Strength of Bolt in Bearing:

$$= \frac{2.5 \times 0.4907 \times 16 \times 12 \times 400}{1.25} = 75372 \text{ N} > 7.66121 \text{ N.}$$

Hence the assumption that bearing strength is more than deign shear is correct.

Design strength of joint per 40 mm width = 666121 N.

Design strength of solid plate per 40 mm width,

$$= \frac{250 \times 40 \times 12}{1.1} = 109091 \text{ N}$$

Maximum efficiency of joint $\frac{66121}{109091} \times 100.$

$$= 60.61 \text{ \%.}$$

2

Tension Members

2.1 Types of Sections

Type of Steel Section Used As Tension Member

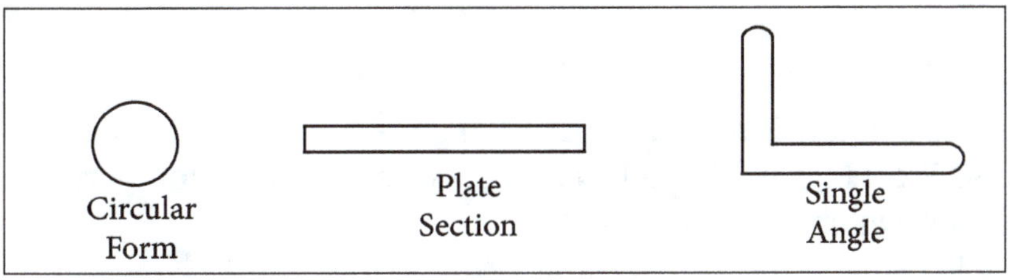

The tension members may be made of single structural shapes. The standard structural shapes of typical tension members are:

- Angle section.
- Tee section.
- Channel section.
- Box section.
- I section.
- Tubular section.

The sections may be built up using a number of top of structural shapes. Single angle members are economical however the connection produces eccentric force within the member. These are usually used in towers and in trusses. Double angle members are more rigid than single angle members. They are used in roof trusses. Since there exists a gap of about 6 to 10 mm between the two members, they are generally interconnected at regular intervals so that they act as one integral member. Within the members of bridge trusses the tensile forces developed are very large and hence require more rigid members. In these structures single channel, single I-section, built-up channels, or built-up I-sections are usually used.

Tension member is the lowest of the following:

- Design strength due to yielding of gross section T_{dg}.

- Rupture strength of critical section, T_{dn}.
- The block shear tab.

Types of tension steel members:

- Carbon Steel.
- High Strength Carbon Steel.
- Medium and High Strength Micro Alloyed Steel.
- High Strength Quenched and Tampered Steels.
- Weathering Steels.

2.1.1 Net Area

The performance of a tension member is often governed by the response of its connections. The AISC Steel Manual introduces a measure of connection performance known as joint efficiency, which is a function of:

- Material properties (ductility).
- Fastener spacing.
- Stress concentrations.
- Shear lag (Most important of the four and addressed specifically by the AISC Steel Manual).

The AISC Steel Manual introduces the concept of effective net area to account for shear lag effects:

- For bolted connections: $A_e = UA_n$.
- For welded connections: $A_e = UA_g$.

Where,

$$U = 1 - \bar{x}/L \leq 0.9$$

And $\bar{x}$ is the distance from the plane of the connection to the centroid of the connected member and L is the length of the connection in the direction of the load.

2.1.2 Net Effective Sections for Angles and Tee in Tension

The connection has a significant influence on the performance of a tension member. A connection almost always weakens the member and a measure of its influence is called joint efficiency.

Joint efficiency is a function of:

- Material ductility.
- Fastener spacing.
- Stress concentration at holes.
- Fabrication procedure.
- Shear lag.

2.2 Design of Connections in Tension Members

Block Shear Failure

At the connected end, failure of a tension member may occur along a path involving shear along one plane and tension on a perpendicular plane along the fastener and this type of failure is known as block Shear Failure.

Problems

1. Let us discuss a double angle tension member connected on each side of a 10 mm thick gusset plate, to carry an axial factored load of 375 kN. Use 20 mm black bolts. Assume shop connection.

Solution:

Area required from the consideration of yielding,

$$= \frac{1.1 \times 375 \times 1000}{250} = 1650 \text{ mm}^2$$

Try 2 ISA 7550, 8 mm thick which has gross area,

$$= 2 \times 938 = 1876 \text{ mm}^2.$$

- In double shear $= \left[\frac{\pi}{4} \times 20^2 + 0.78 \times \frac{\pi}{4} \times 20^2\right] \times \frac{400}{\sqrt{3}} \frac{1}{1.25}$
- Strength in bearing.

Taking e = 40 mm, p = 60 mm,

K_b is smaller of $\frac{40}{3 \times 22}, \frac{60}{3 \times 22} - 0.25, \frac{400}{410}, 1.0$

i.e.,

$$K_b = 0.606$$

$$V_{dpb} = \frac{1}{1.25} \times 2.5 \times 6.606 \times 20 \times 8 \times 400 = 77568 \text{ N}$$

Bolt value = 77568 N.

Number of bolts required = $\frac{375000}{77568} = 4.83$

Provide 5 bolts in a row as shown in figure:

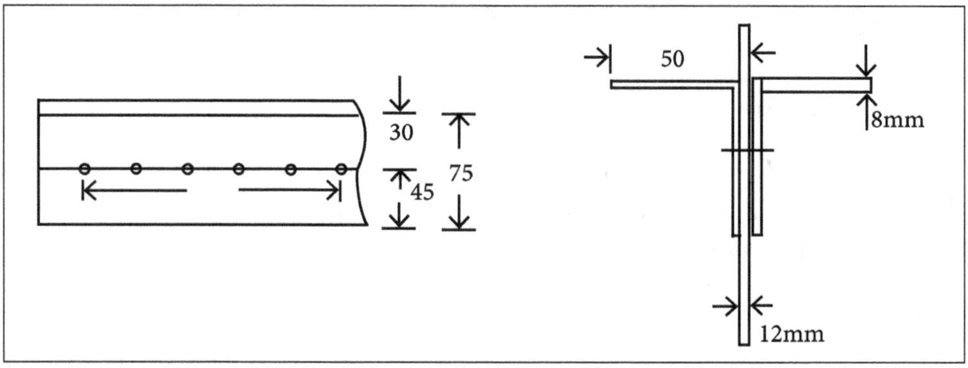

Checking the Design

1. Strength Against Yielding:

$$= \frac{A_g f_v}{\gamma_{mo}} = \frac{1876 \times 250}{1.1} = 426364 \text{ N} > 375 \times 1000$$

2. Strength of Plate in Rupture:

Area of connected leg,

$$A_{nc} = 2\left(75 - 22 - \frac{8}{2}\right) \times 8 = 784 \text{mm}^2.$$

Area of outstanding leg,

$$A_{ge} = 2 \times \left(50 - \frac{8}{2}\right) \times 8$$

$$= 736 \text{ mm}^2$$

$$\beta = 1.4 - 0.076 \times \frac{}{t} \times \frac{}{f_y} \times \frac{}{l_c}$$

$$= 1.4 - 0.076 \times \frac{50}{8} \times \frac{250}{410} \times \frac{77}{240}$$

1.307

$$\therefore \quad _{dn} = \frac{0.9\, f_u\, A_{nc}}{\gamma_{ml}} + \beta \frac{A_{go}\, f_y}{\gamma_{mo}}$$

$$= \frac{0.9 \times 410 \times 784}{1.25} + 1.307$$

$$\frac{736 \times 250}{1.1}$$

$$= 450062 > 175000 \text{ H.}$$

3. Strength against block shear failure:

Per angle,

$$A_{vg} = (40 + 60 \times 4) \times 8 = 2240 \text{ mm}^2.$$

$$A_{vn} = (40 + 60 \times 4 - 4.5 \times 22) \times 8 = 1448 \text{ mm}^2.$$

$$A_{tg} = (75 - 35) \times 8 = 320 \text{ mm}^2.$$

$$A_{in} = (75 - 35 - 0.5 \times 22) \times 8 = 232 \text{ mm}^2.$$

2. Let us determine the tensile load carrying capacity of 2ISA 75 × 75 × 8mm placed back to back of a 10mm thick gusset plate using a single row of 4 nos of 16mm diameter bolts at distance of 40mm from the toe of the angle. Take pitch = 50mm and end distance = 30mm. The length of the member is 4m. Use Fe410 grade steel.

Solution:

Tensile Load Carrying Capacity:

- 2 ISA.
- 75 × 75 × 8mm.
- Connected to 10mmlk gusset plate using single row of 4 Nos. of ϕ 16mm at a distance of 40mm from toe of angle.
- Pitch = 50 mm.

- End distance = 30 mm.
- Length of member = 4 m.
- Use F3410 grade steel.

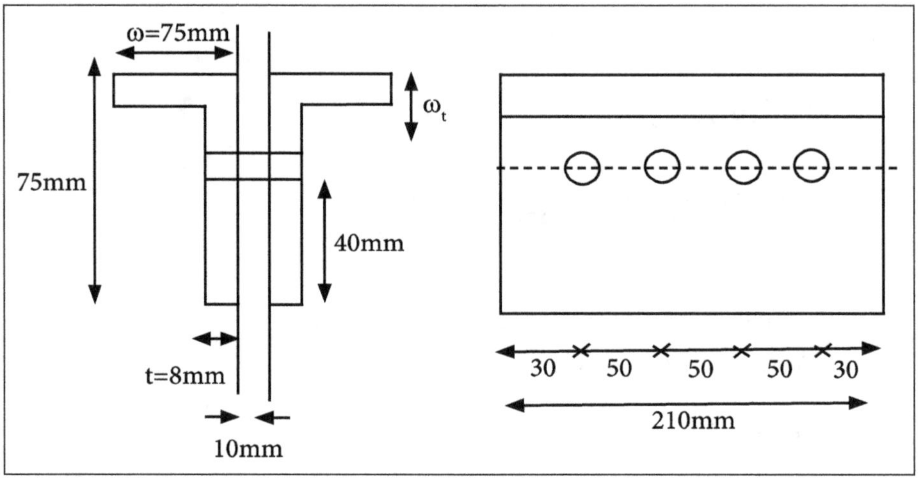

Gross Area of Angles = 2 × 1138 = 2276 mm².

(i) Strength against Yielding,

$$T_{dg} = \frac{A_g f_y}{\gamma_{m_o}} = \frac{2276 \times 250}{1.1} = 517272 \text{ N.}$$

(ii) Strength against Rupture,

Net area connected leg,

$$A_{nc} = \left[75 - \frac{08}{2} - 18\right] \times 8 \times 2$$
$$= 7848 \text{ nun}^2$$

Gross area outstanding leg,

$$A_{ng} = 2\left[75 - \frac{8}{2}\right]8$$
$$= 1136 \text{ mm}^2.$$

$$b_S = [\omega + \omega_t \times t]$$
$$= 75 + 35 - 8.$$

$b_s = 102$ mm.

$L_c = 3 \times 50 = 150$ mm.

$$\beta = 1.4 - 0.076\ [w/t]\ [fy/fu]\ [b_s / L_C] \leq \left[\frac{f_u \gamma_{mo}}{f_y \gamma_{m1}}\right] \geq 0.7$$

[as per Cl. 6.33 of IS 800:2007],

$$= t.4 - 0.076\ [75/8] \left[\frac{250}{410}\right]\left[\frac{102}{150}\right]$$

$$\beta = 1.105$$

$$T_{dn} = \frac{0.9 \times A_{nc} \times f_u}{\gamma_{m1}} + \frac{\beta\ A_{go} f_y}{\gamma_{mo}}$$

$$= \frac{0.9 \times 9848 \times 410}{1.25} + \frac{1.105 \times 1136 \times 250}{1.1}$$

$$T_{dh} = 575$$

(iii) Strength against Block Shear,

$$A_{vg} = [[30 + (3 \times 50)]] \times 8 = 1440\,mm^2$$

$$A_{un} = [30 + (3 \times 50) - 3.5 \times 18] \times 8 = 936\,mm^2.$$

$$A_{tg} = [75 - 35]\ [8] = 3200\ mm^2.$$

$$A_{tn} = [75 - 35 - (0.5 \times 18)]\ [8] = 248\ mm^2.$$

$$T_{db1} = \frac{A_{vg} f_y}{\sqrt{3}\gamma_{mo}} + \frac{0.9 A_{tn} f_u}{\gamma_{m1}}$$

$$= \frac{1440 \times 250}{\sqrt{3} \times 1.1} + \frac{0.9 \times 248 \times 410}{1.25}$$

$$T_{db1} = 262.16\ kN.$$

$$T_{db2} = \frac{0.2 A_{vn} f_u}{\sqrt{3}\gamma_{m1}} + \frac{A_{tg} f_y}{\gamma_{mo}}$$

$$= \frac{0.9 \times 936 \times 410}{\sqrt{3} \times 1.25} + \frac{320 \times 250}{1.1}$$

$T_{db2} = 232.25\,kN.$

T_{db} for Single Angle = 232.25 kN.

T_{db} for 2 Angle = 2 × 232.25

= 464.5 kN.

∴ The design Tensile Load carrying is, 464.5 kN.

3. Let us design a tension member to carry a factored tensile load of 300 kN. The 3m long tension member is connected to a gusset plate 16mm thick with one line of 20mm diameter bolts of grade 4.6. Use Fe410 grade steel.

Solution:

Design Tension Member:

- Factored tensile Load = 300 kN.
- Length of member = 3 m.
- Gusset plate = 16 mmtk.
- Diameter of bolts = 20 mm.
- Grade of bolts = 4.6.
- Use Fe410 grade steel.

$$\Rightarrow Tu = \frac{A_g f_y}{\gamma_{m_0}}$$

$$Ag\ [required] = \frac{T_u\,\gamma_{mo}}{fy}$$

$$= \frac{300 \times 10^3 \times 1.1}{250}$$

$$A_g = 1320\ mm^2.$$

Consider a 2 Angles with long leg outstanding and referring Steel Table, let us choose two Angles ISA 80 × 50 × 6 mm with a gross Area = $1492mm^2$.

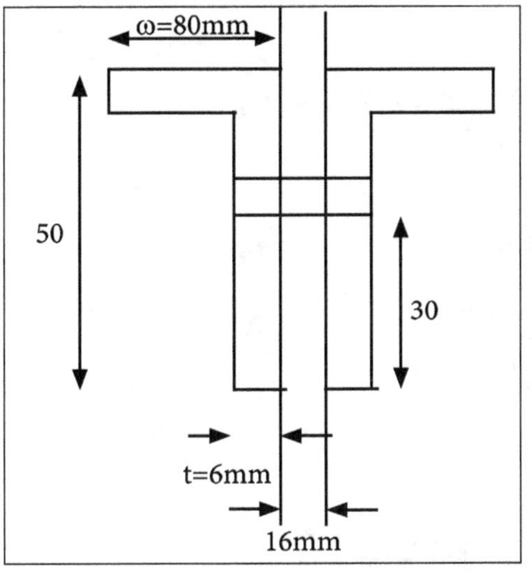

1. Strength in Double Shear:

$$=\frac{f_{ub}}{\sqrt{3}}\frac{1}{\gamma_{m_o}}\left[\frac{\pi}{4}[0.78]d^2+\frac{\pi}{4}d^2\right]$$

$$=\frac{400}{\sqrt{3}}\times\frac{1}{1.25}\left[\frac{\pi}{4}0.78\times20^2+\frac{\pi}{4}20^2\right]$$

$$=\frac{400}{\sqrt{3}}\times\frac{1}{1.25}\left[\frac{\pi}{4}0.78\times20^2+\frac{\pi}{4}20^2\right]$$

$$=103265\,N.$$

2. Strength in Bearing, Vdpt:

Assume e = 30 mm; P = 50 mm.

$$K_b \text{ is least of } \frac{e}{3d_o},\frac{P}{3d_o}-0.25,\frac{f_{yt}}{f_u},1$$

$$=\frac{30}{3\times22},\frac{50}{3\times22}-0.25,\frac{400}{410},1$$

0.4545, 0.5076, 0.9756, 1

$$K_6=0.4545.$$

$$V_{dpb}=\frac{2.5\times K_b dt\ f_u}{\gamma m_o}$$

$$=\frac{2.5\times0.4545\times20\times6\times410}{1.25}=44722\ N.$$

Bolt Value = 44722 N.

No. of bolt required = $\dfrac{300000}{44722} = 6.7 \approx 7$ Nos.

3. Checking the Design:

(i) Strength against Yielding,

$$T_{dg} = \frac{A_g f_y}{\gamma_{m1}} = \frac{1492 \times 250}{1.1}$$

$$T_{dg} = 339091 \text{ N} > 300000 \text{ N},$$

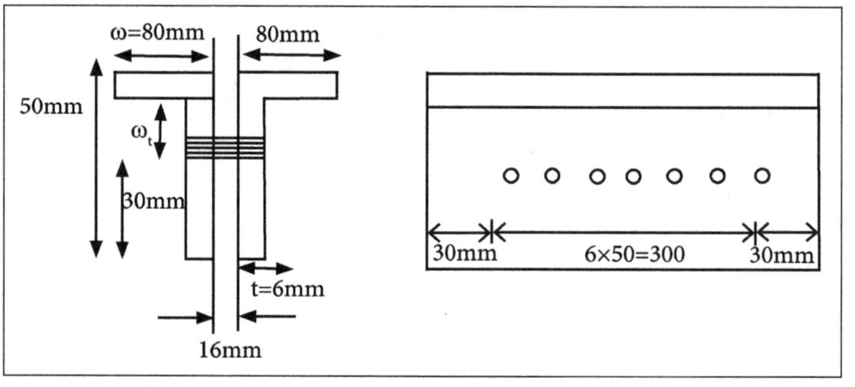

(ii) Strength against Rupture,

Net area of connected leg,

$$A_{nc} = 2\left[50 - \frac{6}{2} - 22\right]6 = 300 \text{ mm}^2.$$

Gross area of outstanding leg,

$$A_{go} = 2\left[80 - \frac{6}{2}\right] \times 6 \times 2 = 924 \text{ mm}^2.$$

$$\omega = 80 \text{ mm},\ t = 6 \text{ mm},\ b = [80 + 20 - 6] = 94 \text{ mm}.$$

$$L_c = 6 \times 50 = 300 \text{ mm}.$$

$$\beta = 1.4 - 0.076\left[\frac{80}{6}\right]\left[\frac{250}{410}\right]\left[\frac{94}{300}\right] \le 1.4432 \ge 0.7$$

$$= 1.2064 \le 1.4432 \ge 0.7$$

$\beta = 1.2064.$

$$T_{dn} = \frac{0.9[410][300]}{1.25} + \frac{1.2064 \times 924 \times 250}{1.1}$$

$= 341904 \text{ N } 300000\text{N}.$

(iii) Strength against Block Shear,

$$A_{vg} = [30 + (6 \times 50)] \times 6 = 1980 \text{ mm}^2.$$

$$A_{vn} = [30 + (6 \times 50) - 6.5 \times 22]\ [6] = 1122 \text{ mm}^2.$$

$$A_{tg} = [50 - 20]\ 6 = 180 \text{ mm}^2.$$

$$A_{tn} = [50 - 20 - 0.5 \times 22] \times 6 = 114 \text{mm}^2.$$

$$T_{db1} = \frac{A_{vg} f_y}{\sqrt{3}\gamma_{mo}} + \frac{0.9 A_{th} f_u}{\gamma_{m1}}$$

$$T_{db1} = \frac{1980}{\sqrt{3}} \times \frac{[250]}{1.1} + \frac{0.9 \times 114 \times 410}{1.25}$$

$$T_{db1} = 293468 \text{ N}.$$

$$T_{db2} = \frac{A_{vn} f_u \times 0.9}{\sqrt{3} \times \gamma_{m1}} + \frac{A_{tg} f_y}{\gamma_{mo}}$$

$$= \frac{0.9 \times 1122 \times 410}{\sqrt{3} \times 1.25} + \frac{180 \times 250}{1.1}.$$

$$T_{db2} = 232141 \text{ N}.$$

$T_{db} \Rightarrow$ form Single Angle = 232141 for Double Angle = 2 × 232141.

T_{db} = 464282 N > 300000 N.

Hence the design is safe.

4. A double angle ISA 75 MM × 75 MM × 8 mm back to back welded to one side 12 mm gusset have allowable stress 150 MPa. Let us determine allowable tensile load on the member and weld length and overlap length of gusset plate.

Solution:

Gross area of angle ISA75 × 75 × 8 mm,

$= 2 \times 1138 = 2276\ \text{mm}^2$

$$A_1 = \frac{A}{2} = \frac{1138}{2} = 569\,\text{mm}^2$$

Effective area,

$$k = \frac{A_1}{A_1 + \frac{A_2}{5}} = \frac{569}{569 + \frac{569}{5}}$$

$= 0.833.$

Effective area,

$= 2\,(A_1 + KA_2)$

$= 2[569 + 0.833\,(569)$

$= 2086\ \text{mm}^2.$

Allowable stress in tension = 150 MPa

Maximum allowable tensile load,

= Effective area × Permissible stress

= 2086 (150)

= 312900 N

= 312.9 kN

Thickness of angle t = 8 mm

Size of Thickness of Well:

(i) Maximum size of weld $= \frac{3}{4}$ (thickness of angle)

$$= \frac{3}{4}(8) = 6\,\text{mm}$$

Thickness thin plate – 1.5 mm

= 0 – 1.5 = 6.5 mm.

(ii) Maximum size = 3 m

= 0.7 (Size of weld)

= 0.7(5) = 3.5 mm

$$f_{wd} = \frac{f_{wn}}{\gamma_{mw}} = \frac{f_u/\sqrt{3}}{\gamma_{mw}}.$$

Assume:

$$f_u = 410 \text{ MPa}, \gamma_{mw} = 1.25$$

$$f_{wd} = \frac{410/\sqrt{3}}{1.25} = 189.39 \text{ Mpa}$$

$$\therefore \text{ Strength of weld } = \frac{1(0.7\times 5)}{1.25}\left(\frac{410}{\sqrt{3}}\right)$$

$$= 662.82 \text{ N} \approx 662 \text{ N}.$$

∴ Length of weld required

$$= \frac{\text{Force}}{\text{Strength}}$$

$$= \frac{312900}{662} = 472.66 \text{ mm}$$

$$= 480 \text{ mm}.$$

$$\therefore \text{ Length of weld } = \frac{486}{2} = 240 \text{ mm}$$

$$\text{Overlap length required } = \frac{240 - 75}{2}$$

$$= 82.5 \text{ say } 85 \text{ mm}.$$

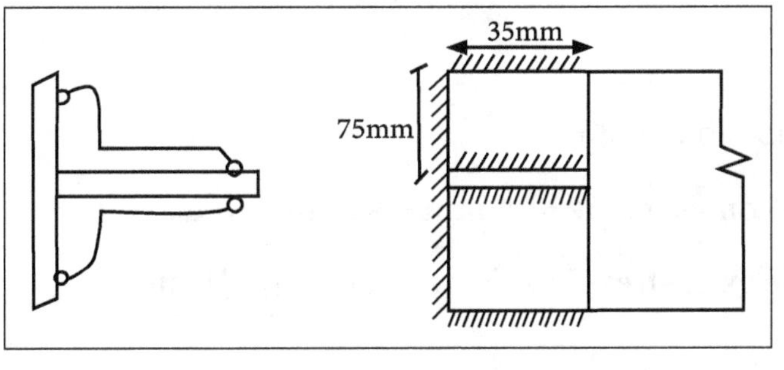

5. An ISA 100 mm × 100 mm × 12 mm is used as a tin riveted to a gusset plate with 24 mm rivets arranged in one row along the length of the angle. Let us determine the allowable tension on the angle is the allowable tensile stress is 150 MPa.

Solution:

Answer: Diameter of the rivet hole = 24 + 1.5 = 25.5 mm,

$$A_{nei} = A_{ess} = A_1 + A_{2k}$$
$$k = \frac{3A_1}{3A_1 + A_2}$$
$$A_1 = A_2 = \frac{A}{2}$$

Grass area $100 \times 100 \times 12$ mm $= 2259$ mm^2,

$$A_1 = A_2 = \frac{2259}{2} = 1129.5 \text{ mm}^2$$
$$k = \frac{3(1129.5)}{3(1129.5) + 1129.5}$$
$$= 0.75$$

$$A_{net} = A_{ess} = A_1 + A_{2k}$$
$$= 1129.5 + 1129.5\ (0.75)$$
$$= 1976.625 \text{ mm}^2.$$

Allowable tension = Allowable Tensile Stress × A_{stress}

$$= 150\ (1976.625)$$

$$= 296493 \text{ N} \approx 296.493 \text{ kN}.$$

6. An angle section 500 × 30 × 6 mm is used as a tension member with its longer leg connected by 12 mm diameter rivets. Let us calculate its strength and take permissible stress in axial tension as l50N/mm².

Solution:

- Angle = 50 × 30 × 6 mm.
- Longer leg connected by = 12 mm diameter rivets.
- Permissible axial stress in axial tension = 150 N/mm².

Rivet:

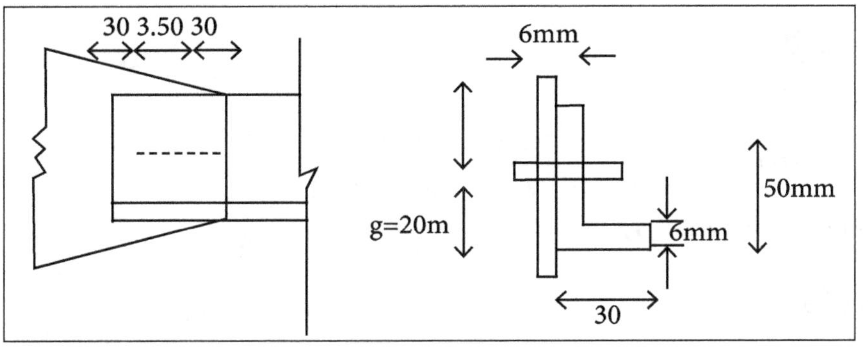

Assume, g = 20mm.

Strength Governed by Yielding:

$$T_{dg} = \frac{A_g f_y}{\gamma_{m_o}}$$

$$A_g = \left[50 - \frac{6}{2} + 30 - 6/2\right] \times 6$$

$$= 0444 \text{ mm}^2$$

$$T_{dg} = \frac{0444 \times 250}{1.1} = 100909.0 \text{ N.}$$

Strength Governed by Tearing at Critical Sections:

Refer CI.6.3.3 of IS800-2007.

$$T_{dn} = \left[\frac{0.9 \times A_{nc} f_u}{\gamma_{m1}}\right] + \left[\frac{\beta A_{go} f_y}{\gamma_{mo}}\right]$$

$$A_{nc} = \left[50 - \frac{6}{2} - 12\right] 6 = 210 \text{ mm}^2$$

$$A_{go} = [30 - 6/2]6 = 162 \text{ mm}^2.$$

$$\beta = 1.4 - 0.076 \left[\frac{w}{t}\right]\left[\frac{f_y}{f_u}\right]\left[\frac{b_s}{L_c}\right] \leq \frac{f_u \gamma_{m_o}}{f_y \gamma_{m_1}} \geq 0.7$$

$$\omega = 30 \text{ mm}$$

$$\omega_t = 20 \text{ mm} ; t = 6 \text{ mm.}$$

$b_s = 30+20-6=44$ mm.

$L_c = 3 \times 50 = 150$ mm.

$$\beta = 1.4 - 0.076\left[\frac{30}{20}\right]\left[\frac{250}{150}\right]\left[\frac{44}{150}\right] \le \frac{440 \times 1.1}{250 \times 1.25} \ge 0.7$$

$$= 1.34 \le 1.443 \ge 0.7$$

$\beta = 1.34$

$$T_{dn} = \frac{0.9 \times 210 \times 410}{1.25} + \frac{1.34 \times 162 \times 250}{1.1}$$

$$= 61992 + 49336 = 111328 \text{ N.}$$

Block Shear Strength:

$$T_{db1} = \frac{A_{vg} f_y}{\sqrt{3}\gamma_{m_o}} + \frac{0.9 A_{tn} f_u}{\gamma_{m_1}}$$

$A_{vg} = [30+[3\times50]]\ 6 = 1080 \text{ mm}^2$

$A_{vn} = [30+(3\times50)-3-5[12]]\times6 = 828 \text{ mm}^2$

$A_{fg} = [50-20]\times 6 = 180 \text{ mm}^2$

$A_{tn} = [50-20-0.5\times12]6 = 144 \text{ mm}^2$.

$$T_{db1} = \frac{1080\times250}{\sqrt{3}\times1.1} + \frac{0.9\times144\times490}{1.25} = 184222 \text{ N.}$$

$$T_{db2} = \frac{A_{tg} f_y}{\gamma_{m_o}} + \frac{0.9 A_{vn} f_u}{\sqrt{3}\gamma_{m_1}}$$

$$= \frac{180\times250}{1.1} + \frac{0.9\times828\times410}{\sqrt{3}\times1.25} = 182028 \text{ N.}$$

$T_{db} = 182028$ N.

Strength of member = 100909N.

If Welds:

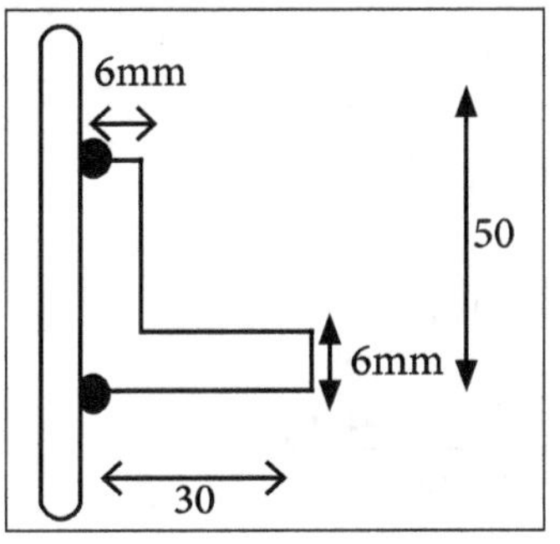

$$A_g = 447 \text{ mm}^2$$

$$A_{nc} = \left[50 - \frac{6}{2}\right]6 = 282 \text{ mm}^2$$

$$A_{go} = \left[30 - \frac{6}{2}\right]6 = 162 \text{ mm}^2$$

Strength Governed by Yielding:

$$T_{dg} = \frac{A_g f_y}{\gamma_{m_0}} = \frac{447 \times 250}{1.1} = 101590 \text{ N.}$$

Strength of Plate from Rupture:

$$T_{dn} = \frac{0.9\, f_u A_{nc}}{\gamma_{m_1}} + \frac{\beta A_{go} f_y}{\gamma_{m_0}}$$

$$\beta = 1.4 - 0.076\left[\frac{w}{t}\right]\left[\frac{f_y}{f_u}\right]\left[\frac{b_s}{L_c}\right] \leq \frac{f_u}{f_y}\frac{\gamma_{m_0}}{\gamma_{m_1}} \geq 0.$$

$$\omega = 30 \text{ mm}_t = 6 \text{ mmb}_s = 30 \text{ mmL}_c = [210] \text{ mm.}$$

$$\beta = 1.4 - 0.076\left[\frac{30}{6}\right]\left[\frac{250}{210}\right] \leq \left[\frac{410}{250} \times \frac{1.1}{1.25}\right] \geq 0$$

$$= 1.36 \leq 1.44 \geq 0.7\beta = 1.36.$$

$$T_{dn} = \frac{0.9 \times 410 \times 282}{1.25} + \frac{1.36 \times 162 \times 250}{1.1}$$

$$= 133319 \text{ N.}$$

Block Shear Strength:

Assume gusset plate = 10 mm.

$$A_{vg} = 2\left[\frac{210+210}{2}\right] \times 10 = 4200\,mm^2.$$

$$A_{vg} = A_{vn} = 4200\ mm^2$$

$$A_{vg} = A_{tn} = 50 \times 10 = 500\ mm\ .$$

$$T_{db1} = \frac{4200 \times 250}{\sqrt{3} \times 1.1} + \frac{0.9 \times 500 \times 410}{1.25}$$

$$T_{db1} = 698707\ N.$$

$$T_{db2} = \frac{0.9 \times 4200 \times 410}{\sqrt{3} \times 1.25} + \frac{500 \times 250}{1.1}$$

$$T_{db2} = 829458\ N.$$

$$T_{db} = 698707 N.$$

Strength of tension member = 101590 N.

7. Let us design a tension member of length 3.6 m between c/c of intersections and carrying a pull of 150 kN. The member is subjected to reversal of stresses.

Solution:

Gross area of the angle required,

$$T_u = \frac{A_g f_u}{\gamma_{mo}}$$

$$A_g = \frac{150 \times 1000 \times 1.1}{250}$$

$$= 880\ mm_2.$$

Use ISA 100 65, 8 mm with gross area A_g = 1257 mm².

Let us use 10 mm gusset plate for connection.

Diameter of bolt = 20 mm

Diameter of bolt hole = 22 mm.

Strength of one bolt in single shear,

$$V_{dsb} = \frac{f_{ub}}{\sqrt{3}}\left(\frac{n_n A_{nb} + n_s^{\cdot} A_{sb}}{\gamma_{mb}}\right)$$

$$= \frac{400}{\sqrt{3}}\left(\frac{1 \times 0.78 \times \pi \times \frac{20^2}{4}}{1.25}\right)$$

Where,

$= 45251$ N

$= 45.25$ kN.

Adopts edge distances e = 40 mm, pitch P = 60 mm.

$k_b = 0.6061$

$$V_{dpb} = \frac{2.5\, k_b\, dt\, f_u}{\gamma_{mb}}$$

$$= \frac{2.5(0.6061)(20)(8)(410)}{1.25}$$

$= 79520$ N

$= 79.52$ kN.

Bolt Value $= 45251$ N

$= 45.251$ kN.

$$\text{Number of bolts} = \frac{(150)(1000)}{45251} = 5 \text{ bolts}$$

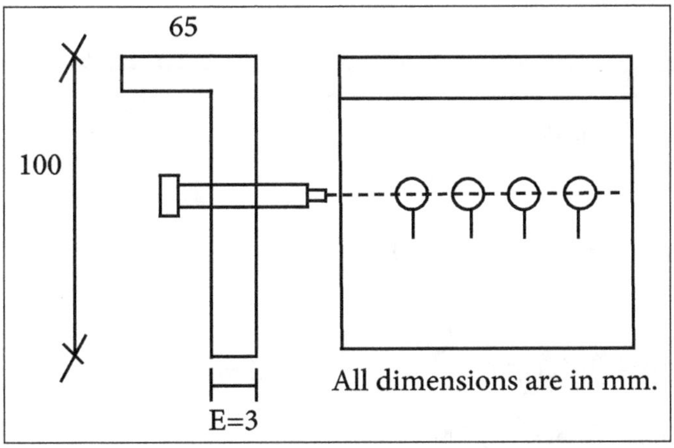

Checking the Design

1. Strength from Yielding Consideration:

$$\frac{A_g \, fy}{\gamma_{mo}} = \frac{1257\,(250)}{1.1}$$

$$= 285.682 \text{ N} > 20000 \text{ N.}$$

2. Strength from Rupture of Critical Section:

Net area of connected leg $A_{nc} = (100 - 22 - 8/2) \times 8$

$$= 592 \text{ mm}^2.$$

Gross area of outstanding leg $= \left(65 - \frac{8}{2}\right) \times 8$

$$= 488 \text{ mm}^2.$$

To calculate β

$W = 65$ mm, $t = 8$ mm, $b_s = (65 + 40 - 8) = 97$ mm

$L_c \Rightarrow$ Length of member between outer most.

Fasteners = 4(60) = 240 mm,

$$\beta = 1.4 - 0.076\left(\frac{W}{t}\right)\left(\frac{fy}{fu}\right)(bs_{lc}) \le \frac{fu}{fy}\frac{\gamma_{mo}}{\gamma_{me}} \ge 0$$

$$\beta = 1.4 - 0.076\left(\frac{65}{8}\right)\left(\frac{250}{410}\right)\left(\frac{97}{240}\right)$$

$$= 1.243 \le 1.4433 \ge 0.7$$

$$\beta = 1.248$$

$$T_{dn} = \frac{0.9\,fu\,A_{nc}}{\gamma_{x\lambda}} + \frac{\beta\,A_{go}\,fy}{\gamma_{mo}}$$

$$= \frac{0.9\,(410)\,(592)}{1.25} + \frac{1.248\,(488)(25)}{1.1}$$

$$= 174750 + 138415$$

$$= 313173 \text{ N} > 200000 \text{ N.}$$

3. From Block Shear Consideration:

$$A_{vg} = [40 + 60(4)]\ (8) = 2240\ mm^2$$

$$A_{vn} = [40 + 60(4) - 4.5(22)]\ (8) = 1448\ mm^2$$

$$A_{fg} = [100 - 40](8) = 480\ mm^2$$

$$A_{tn} = [100 - 40 - 0.5(22)]\ (8) = 392\ mm^2.$$

$$T_{db1} = \frac{A_{vg} f_y}{\sqrt{3}\ \gamma_{mo}} + \frac{0.9\ A_{tn}\ fy}{\gamma_{me}}$$

$$= \frac{2240\ (250)}{\sqrt{3}(1.1)} + \frac{0.9\ (392)(410)}{1.25} = 409642\ N$$

$$T_{db2} = \frac{0.9\ A_{vn}\ fu}{\sqrt{3}\gamma_{me}} + \frac{A_{vg}\ fy}{\gamma_{mo}}$$

$$= \frac{0.9(1498)(410)}{\sqrt{3}(1.25)} + \frac{480\ (250)}{1.1}$$

$$= 355886\ N.$$

$$T_{db} = 355886\ N > 200000\ N.$$

Hence the design is safe.

Check for maximum 1/r value.

Referring to steel tables γ_{min} for angle section

$$100 \times 65 \times 8\ mm = 13.9\ mm$$

$$\frac{1}{\gamma_{in}} = \frac{3000}{13.9} = 215.8 < 350.$$

8. Here we design a tension member to carry a load 300 kN. The two angle placed back to long with long leg outstanding are desirable. The length of the member 3 m.

Solution:

Area required from yielding consideration,

$$T_u = \frac{A_g fy}{\gamma_{mo}} \Rightarrow A_{g\ required} \frac{T_u \gamma_{mo}}{f_y}$$

$$A_{q\ required} = \frac{300(1000)(1.1)}{250} = 1320\,mm^2$$

1. In Double Shear:

$$= \frac{400}{\sqrt{3}} \cdot \frac{1}{1.25}\left[\frac{\pi}{4}(0.78)\times 20^2 + \frac{\pi}{4}\times 20^2\right]$$

$$= 103365\ N.$$

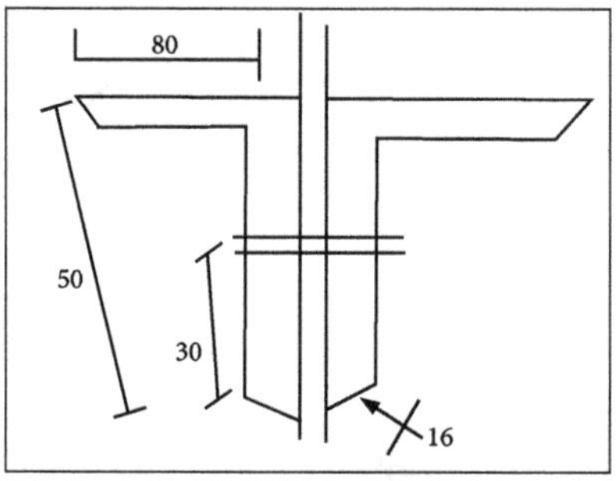

2. Strength in Bearing:

Take e = 30 mm, P = 50 mm

kb is least of $\frac{30}{3(22)} \cdot \frac{50}{3(22)} - 0.25 ; \frac{400}{410} : 1.0$

least of 0.4545; 0.5076; 0.9756; 10

$k_b = 0.4545.$

Strength in bearing,

$$V_{dpb} = \frac{2.5 k_b\, dt\ fu}{\gamma_{mb}}$$

$$= \frac{2.5\,(0.4545)(20)(6)(410)}{1.25}$$

$$= 44722\ N.$$

$\therefore$ Bolt Value = 44722 N.

No. of bolts required = $\frac{300000}{44722} = 6.7 \approx 7$ bolts.

(i) Strength Against Design:

$$= \frac{A_g \cdot fy}{1.1} = \frac{1492\,(250)}{1.1}$$

$$= 33909 \text{ N} > 300000 \text{ N}.$$

(ii) Strength against rupture:

$$\text{Net area of connected Anc} = 2\left(50 - \frac{6}{2} - 22\right)(6)$$

$$= 300 \text{ mm}^2.$$

$$A_{go} = 2\left(20 - \frac{6}{2}\right)(6)(2) = 924 \text{ mm}^2.$$

$$W = 80 \text{ mm}, \; t = 6 \text{ mm}.$$

$$b_s \; (80 + 20 - 6) = 94 \text{ mm}, \; L_C = 6(50) = 300 \text{ mm}$$

$$\beta = 1.4 \; - \; 0.076\left(\frac{80}{6}\right)\left(\frac{25}{410}\right)\left(\frac{94}{300}\right)$$

$$\beta = 1.2064$$

$$T_{du} = \frac{0.9(410)(300)}{1.25} + \frac{1.2064\,(924)(250)}{1.1}$$

$$= 3419.4 > 300000 \text{ N}.$$

(iii) Strength Against Block Shear:

$$A_{vg} = \left[30 \; + \; (6 \times 50)\right] \; (6) = 1980 \text{ mm}^2$$

$$A_{vn} = \left[30 \; + \; 6 \times 50 \; - \; 6.5(22)\right] \; (6) = 1122 \text{ mm}^2$$

$$A_{tg} = [50 \; - \; 20](6) = 180 \text{ mm}^2$$

$$A_{tn} = \left[50 \; - \; 20 \; - 05 \; (22)\right] \; (6) = 114 \text{ mm}^2.$$

Least Value T_{db1} and T_{db2},

Where,

$$T_{db1} = \frac{A_{vg}\, fy}{\sqrt{\gamma_{mo}}} + \frac{0.9\, A_{tn}\, fu}{\gamma_{m1}}$$

$$= \frac{1980\,(250)}{\sqrt{3}\,(1.1)} + \frac{0.9(114)(410)}{1.25}$$

$$= 293468 \text{ N.}$$

$$T_{db2} = \frac{0.9(1122)(410)}{\sqrt{3}\,(1.25)} + \frac{180(250)}{1.1}$$

$$= 237141 \text{ N.}$$

T_{db} for single angle = 232141 N

T_{db} for 2 angle = 2 × 232141 = 464782 N

C = 464282 N > 300000 N

Check for maximum $\frac{l}{\gamma_{min}}$ value $= \frac{2900}{13.9} = 208 < 350.$

9. Here we design a tension member to carry a load of 300 kN. The two angles placed back to back with long leg outstanding are desirable. The length of the member is 2.9 m.

Solution:

Met over required,

$$= \frac{180 \times 10^3}{150} = 1200 \text{ mm}^2.$$

Let us try a single equal angle 90 mm × 90 mm × 90 mm. Net area of the connect leg,

A_1 = (90 – 5)10 – 21.5 × 10 = 635 mm².

Area of the outstanding leg,

A_2 = (90 – 5)10 = 850 mm²

$$K = \frac{3A_1}{3A_1 + A_2} = \frac{3 \times 635}{3 \times 635 + 850} = 0.6915$$

Effective area,

$A_{eff} = A_1 + kA_2$ = 635 + (0.6915 × 850)

A_{eff} = 1222.775 mm².

Safe tension of the member = $\dfrac{1222.775 \times 150}{1000} = 183.42$ kN.

But tension in the member only 180 kN. Therefore design safe.

Let us try a single unequal angle.

100mm × 65 mm × 10 mm

$A_1 = (100 - 5)10 - 21.5 \times 10 = 735 \text{ mm}^2$.

Area of the outstanding leg,

$A_2 = (65 - 5)10 = 600 \text{ mm}^2$

$$K = \frac{3A_1}{3A_1 + A_2} = \frac{3 \times 735}{3 \times 735 + 600} = 0.7861$$

Effective area of the member,

$A_{eff} = A_1 + KA_2 = 735 + 0.7861 \times 600$

$= 1,206.66 \text{mm}^2$

Safe tension for the member $= \dfrac{1206.66 \times 150}{1000} = 180.999$ kN

The design is safe.

2.2.1 Use of Lug Angles

Lug Angles

Lug Angles connecting outstanding leg of a Channel-Shaped member shall, as far as possible be disposed symmetrically with respect to the section of the member.

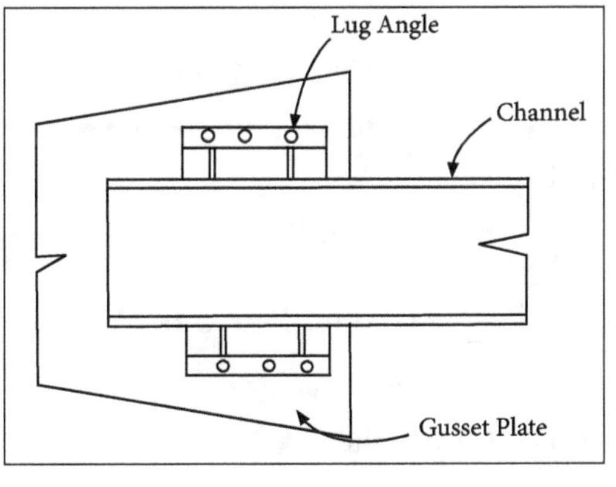

Sometime heavily loaded tension member may require large length of end connection. In order to reduce such lengthy connection, Lug angle are used. Lug angle are helpful is saving of gusset plate, but will result in use of additional fasteners and angle.

2.3 Design of Tension Splice

Tension Splices

It is a common practice to fabricate members in manageable lengths and then to join them after placing them in their proper locations in the field with suitable connections to form the complete structure. In effect the members are cut at certain chosen points and need to be effectively joined (or "spliced) there. Such a splice in a tension member has the function of replacing the member at such cuts for continuity.

Problems

1. Let us design a splice to connect a 300 × 20 mm plate with a 300 × 10 mm plate. The design load is 500 kN use 20 mm black bolts, fabricated in the shop.

Solution:

Let double cover but joint with 6 mm cover plates be used.

Strength of Bolts:

Strength in double shear,

$$=\beta_{pk}\left(\frac{\pi}{4}d^2+0.78\frac{\pi}{4}d^2\right)\frac{f_c}{\sqrt{3}\times 1.25}$$

$$=0.875\times 1.78\times\left(\frac{\pi\, 20^2}{4}\right)\times\frac{400}{\sqrt{3}}\times\frac{400}{\sqrt{3}}\times\frac{1}{1.25}$$

$$=90400\text{ N.}$$

Strength in Bearing:

Let edge distance = 40 mm and pitch 60 mm be used.

Then Kb is smaller of $\frac{e}{3d_o},\frac{p}{3d_o}-0.25,\frac{f_{ub}}{f_u},1.0,$

i.e.,

$$=\frac{40}{3\times 22},\frac{60}{3\times 22}-0.25,\frac{400}{410},1.0 \quad \text{i.e., } K_b=0.606$$

∴ Strength in bearing against 10 mm plate

$$=2.5K_b\, dtf_u\times\frac{1}{\gamma_{wb}}$$

$$= 2.5 \times 0.606 \times 20 \times 10 \times 410 \times \frac{1}{1.25}$$

$$= 99384 \text{ N } 105897 \text{ N}$$

$\therefore$ Bolt value = 90400 N

Hence number of bolts required = $\frac{500 \times 1000}{90400} = 5{\cdot}53$

Check for the Strength of Plate:

(i) Strength against yielding of gross section:

$$\frac{A_g f_y}{\gamma_{mo}} = \frac{300 \times 10 \times 250}{1.1} = 681818 \text{ N} > 50000 \text{ N.}$$

(ii) Strength against rupture:

$A_n = (300 - 3 \times 22) \times 10 = 2340 \text{ mm}^2$

$$\therefore T_{dm} = \frac{0.9\, A_n\, f_u}{\gamma_{ml}} = \frac{0.9 \times 234 \times 410}{1.25}$$

$$= 690768 \text{ N } 500000 \text{ N.}$$

(iii) Block shear strength:

$$T_{db} = \frac{A_{vg}\, f_y}{\sqrt{3}\,\gamma_{mo}} + \frac{0.9\, A_{tn}\, f_u}{\gamma_{ml}}$$

$$= \frac{2000 \times 250}{\sqrt{3} \times 1.1} + \frac{0.9 \times 1760 \times 410}{1.25}$$

$$T_{ab} = 781984 \text{ N.}$$

2. Here we design a tension splice to connect two plates of size 250 mm × 20 mm and 220 mm × 12 mm, for a design load of 250 kN.

Solution:

Design Tension Splice:

- 250 × 20 mm and 220 × 12 mm.
- Design Load = 250kN.

Let us provide double cover butt joint with 6 mm cover plates on either side with 20 mm diameter black belts.

Strength of Bolts:

$d = 20$ mm, $d_{12} = 22$ mm, $\beta_{pk} = 1 - 0.0125\,[10] = 0.875$.

Strength in double Shear

$$=\beta_{PK}\left[0.78\frac{\pi d^2}{4}+\frac{\pi d^2}{4}\right]\frac{f_u}{\sqrt{3}}\times\frac{1}{1.25}$$

$$=0.875\left[0.78\times\frac{\pi\times 20^2}{4}+\frac{\pi\times 20^2}{4}\right]\times\frac{400}{\sqrt{3}}\times\frac{1}{1.25}$$

$$=90399\text{ N}$$

Strength in Bearing:

Let us provide edge distance, e = 40 mm.

Pitch, P = 60 mm.

$$k_b=\frac{40}{3\times 22};\frac{60}{3\times 22}-0.25;\frac{400}{410};1.0$$

$$k_1=n_o/n_f$$

$\therefore$ Strength in bearing against 12 mm plate.

$$=2.5\times 0.606\times 20\times 12\times 410\times\frac{1}{1.25}$$

Bolt Valve = 90399 N.

Number of bolts required $=\dfrac{250\times 1000}{90399}=2.76\approx 3$bolt.

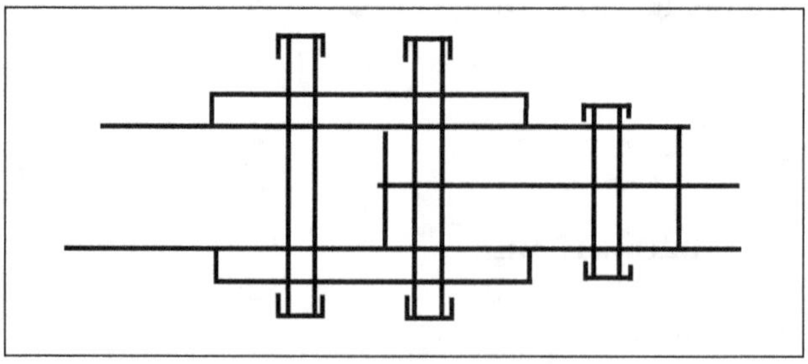

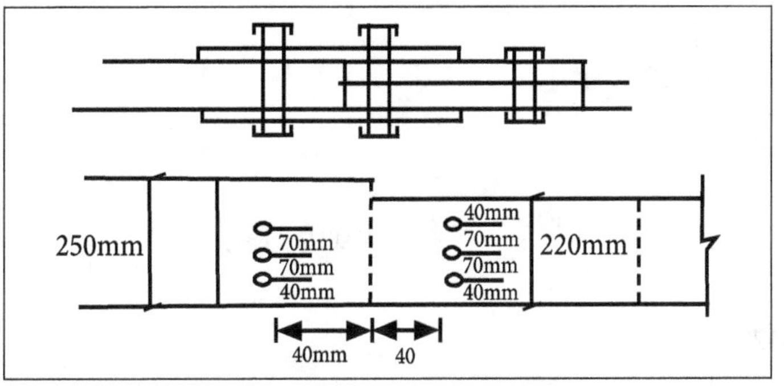

Checking the Design

(i) From Yielding Consideration

Strength against yielding $= \dfrac{[220 \times 12] \times 250}{1.1}$

$= 5600000\ 250000\text{N}.$

(ii) From Rupture Consideration

Strength against rupture,

$$T_{dn} = \frac{0.9\ A_n f_u}{\gamma_{m_1}}$$

$$A_n = [220\ - 3(22)]\ \times\ 12 = 1848\ \text{mm}^2.$$

$$T_{dn} = \frac{0.9 \times 1848 \times 410}{1.25} = 545529 > 250000\ \text{N}$$

(iii) Block Shear Strength

$$A_{vg} = 3\ \times\ 40\ \times 12 = 1440\ \text{mm}^2.$$

$$A_{vn} = 3[[40]\ - 0.5\ \times\ 22]\ 12 = 1044\ \text{mm}^2.$$

$$A_{kg} = [40 + 40]12 = 960\ \text{mm}^2.$$

$$A_{tn} = [40 + 40 - 2\ \times 22] \times 12 = 432\ \text{mm}^2.$$

$$T_{db1} = \frac{A_{vg} f_y}{\sqrt{3} \cdot \gamma_{m_0}} + \frac{0.9 A_{tn}\ f_u}{\gamma_{m1}}$$

$$= \frac{1440 \times 250}{\sqrt{3} \times 1.1} + \frac{0.9 \times 432 \times 410}{1.25}$$

$$T_{db1} = 3.16477 \text{ N}$$

$$T_{db2} = \frac{0.9\, A_{vn}\, f_u}{\sqrt{3}\,\gamma_{m_1}} + \frac{A_{tg}\, f_y}{\gamma_{m_0}}$$

$$= \frac{0.9 \times 1044 \times 410}{\sqrt{3} \times 1.25} + \frac{960 \times 250}{1.1} = 396115 \text{ N}$$

$$T_{db} = 316477 \text{ N}$$

Let us provide 2 number of 30 mm dia meter half with one extra bolt in the caver.

3. Let us discuss the problem based on a tension splice for tension member sections 160 mm × 100 mm and 250 mm × 12 mm. The member is subjected to a pull of 200 kN.

Solution:

Load transmitted to splice plates,

= 50% of the column load

$$= \frac{50}{100} \times 300 = 150 \text{kN}.$$

Load on one splice plate $= \frac{150}{2} \text{N/mm}^2.$

Area of each splice plate $= \frac{75 \times 10^3}{150} = 500 \text{mm}^2.$

Width of splice plate = width off lange of the column,

= 225 mm

Thickness of the splice plate $= \frac{500}{225} = 2.22 \text{mm}$

Provide a thickness of 6mm.

Strength of the riyert in single shear,

$$= f_s \frac{\pi d^2}{4}$$

$$= 100 \times \pi \times \frac{19.5^2}{4}$$

$$= 29.865 \text{ N}$$

Strength of one rivert in bearing = f_b dt,

$$= 300 \times 19.5 \times 6$$

$$= 35, 1000 \text{ N}$$

Lesser rivert value = 25,865 N.

Number of riverts required on each side of the splice section,

$$= \frac{75 \times 10^3}{29865} = 2.6$$

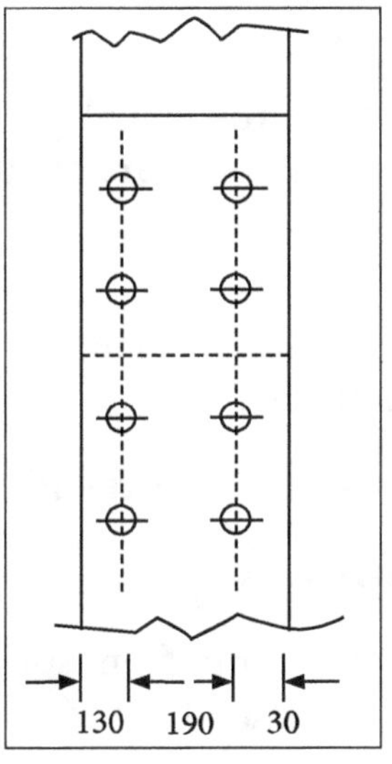

The rivets are arranged in two rows.

2.3.1 Concept of Shear Lag

Block Shear

If the connected end, failure of a tension member may occur along a path involving shear along one lane and tension on a perpendicular plane along the fastener.

Concept of Shear Lag

Shear lag is a phenomenon in which the stiffer regions of the structure or structural component attract more stress than the flexural regions.

3

Compression Members

3.1 Types of Compression Members

Effective Length of a Column

The effective length is calculated from the actual length L, if the member considering the rotational and relative transitional boundary conditions at the ends.

Latticed Columns

Lattice columns are made of parallel supports interlaced with each other by strapping found in diamond or triangular patterns.

The most famous example of lattice column is the Eiffel towers.

Eccentrically Loaded Column

When the load is transferred to the column from some eccentricity (the distance from the assumed point of application of the load to the centroid of column).

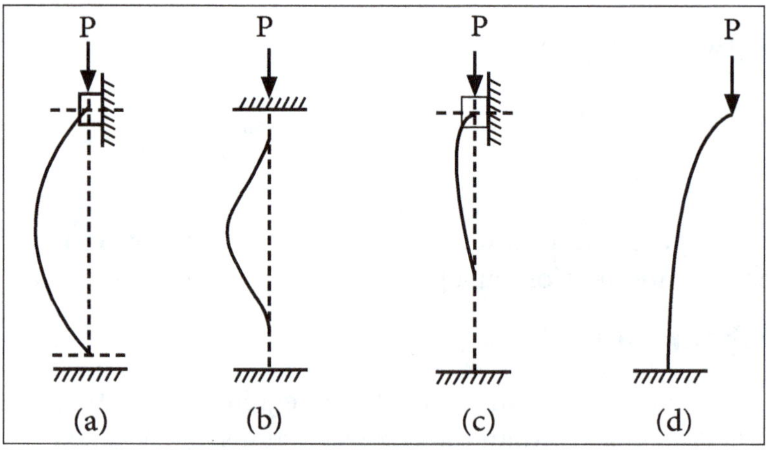

Types of end conditions of columns.

The column are classify as under Euler's condition:

- Both end fixed.
- Both end pinned or hinged.

- One end fixed and other end hinged.
- One end fixed and other end free.

Let us discuss the axial load carrying capacity of a stanchion ISMB 300, 5m high. The column is pinned at both the ends. fy = 250 MPa.

Solution:

Axial Load Carrying Capacity,

- ISMB 300.
- High = 5 m.
- Column → Pinned @ both ends.
- f_y = 250 MPa.

1. Properties of Section

f_y = 250 MPa.	f_u = 410 MPa.
A_e = A = 5626 mm².	E = 2 × 10^5 MPa.

Effective Length [KL] for both ends Pinned,

= 1 x 3000

= 3000 mm [as per Table of IS 800:200]

h = 300 mm

b_f = 140 mm

b_f = 12.4mm

$\gamma_{min} = \gamma_y = 28.4\text{mm}$

2. Buckling Class Determined

[As Per Table of IS 800],

$$\frac{h}{h_f} = \frac{3000}{140} = 2.14 \text{ and } t_f = 12.4 \ [\text{i.e}] \leq 2100\text{mm}.$$

The classification for buckling class is a follows. It falls under buckling class 'b' for buckling about zz axis and 'c' about YY axis.

3. Euler Buckling Stress, f_u

$$f_{cc} = \frac{\pi^2 E}{\left[\frac{KL}{\gamma_{min}}\right]^2}$$

$$= \frac{\pi^2\left[2\times10^5\right]}{\left[\frac{3000}{28.4}\right]^2} = 176.90 \text{ MPa.}$$

4. Calculation of Effective Slenderness Ratio $[\lambda]$

$$\lambda = \sqrt{\frac{f_y}{f_{cc}}} \text{ [as per Cl 7.1.2.1 of IS 800:200].}$$

$$= \sqrt{\frac{250}{176.90}}$$

$$\lambda = 1.18$$

5. Computation of F_d

As per Cl-7.1.2.1 of IS 800:2007,

$$f_d = \frac{f_y/\gamma_{mo}}{\phi + \left[\phi^2 - \lambda^2\right]^{0.5}} \leq \frac{f_y}{\gamma_{mo}}$$

$$\phi = 0.5\left[1 + \alpha\left[\lambda - 0.2\right] + \lambda^2\right]$$

∂ = Imperfection factor [as per Table of IS 800-2007]

= 0.34 'for buckling class 6b' about ZZ-axis.

$$\phi = 0.5[1+0.34][0.18-0.2]+[1.182]$$

$$\phi = 1.36$$

$$f_y = 250\text{MPa.}$$

$$\gamma_{m_o} = 1.10$$

[as per Table of IS 800:2007],

$$f_{cd} = \frac{250/1.1}{01.36 + \left[1.36^2 - 1.18^2\right]^{0.5}} \leq \frac{f_y}{\gamma_{mo}}$$

$$f_{cd} = 111.62 \leq \frac{250}{1.1}$$

$$f_{cd} = 111.62 \text{ MPa.}$$

6. Computation of Strength of Column

$$Pd = A_e f_{cd}$$

$$= 5626 \times 111.62 = 627974 \text{ N.}$$

$$Pd = 627.974 \text{kN.}$$

3.1.1 Theory of Columns

Euler's Theory

The first rational attempt to check the stability of long columns was created by Mr. Euler. He derived an equation for the buckling load of long columns based on the bending stress. Whereas deriving this equation, the effect of direct stress is neglected. This could be justified with the statement that the direct stress induced during a long column is negligible as compared to the bending stress. It should be noted that Euler's formula cannot be utilized in the case of short columns, because the direct stress is appreciable and thus cannot be neglected.

Assumptions in Euler's Column Theory

The following simplifying assumptions are made in Euler's column theory:

- Initially the column is perfectly straight, and the load applied is truly axial.
- The cross-section of the column is uniform throughout its length.
- The column material is perfectly elastic, homogeneous and isotropic, and thus obeys Hooke's law.
- The length of column is very large as compared to its cross-sectional dimensions.
- The shortening of column, due to direct compression (being very small) is neglected.
- The failure of column occurs due to buckling alone.
- The weight of the column itself is neglected.

Euler's Formula

According to Euler's theory, the crippling or buckling load (W_{cr}) under various end conditions is represented by a general equation,

$$W_{cr} = \frac{C\pi^2 EI}{l^2} = \frac{C\pi^2 EAk^2}{l^2} \qquad ...(\because I = A.k^2)$$

$$= \frac{C\pi^2 EA}{(lIk)^2}$$

Where,

E = Modulus of elasticity or Young's modulus for the material of the column.

A = Area of cross-section.

K = Least radius of gyration of the cross-section.

l = Length of the column and

C = Constant, representing the end conditions of the column or end fixity coefficient.

Basic Strut Theory

$$EI\, d^2 y/dx^2 - P y \qquad ...(1)$$

Euler's critical load is obtained as,

$$P_E = \frac{\pi^2 EI_y}{l^2} \qquad ...(2)$$

In terms of the stress equation is,

$$p_E = \frac{\pi^2 E}{(KL/r)^2} \qquad ...(3)$$

Strut with Initial Curvature

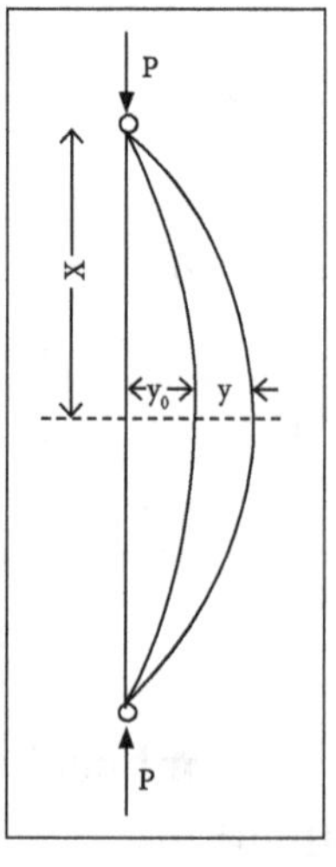

In practice, columns are generally not straight and the effect out of straightness on strength is studied. Consider a strut with an initial curvature bent in a half sine curve as shown in Figure.

If the initial deflection, at x from A is y_o load P the equilibrium equation is,

$$EI\frac{d^2y}{dx^2}=P(y+y_o) \qquad ...(4)$$

Where deflection $y=\sin\left(\frac{\pi x}{l}\right)$ ...(5)

If δ_o is the deflection at the centre and δ the additional deflection caused by p.

Then,

$$\delta=\frac{\delta_o}{(P_E/P)-1} \qquad ...(6)$$

The maximum stress at the centre of the strut is given by,

$$P_{max}=\frac{P}{A}+\frac{P(\delta_o+\delta)h}{I_y} \qquad ...(7)$$

Where h is shown in figure,

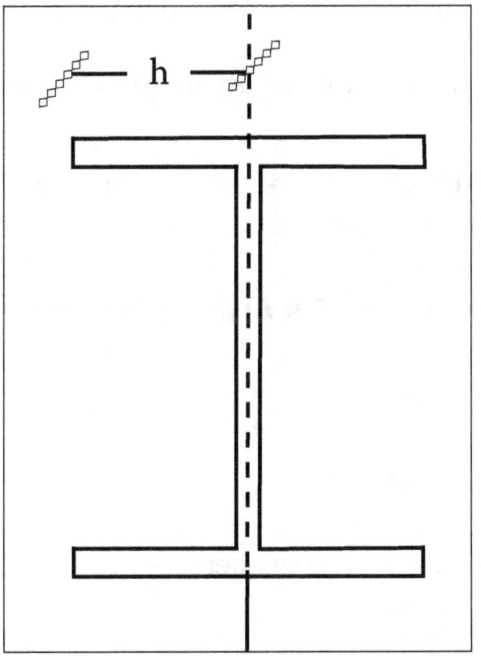

i.e.,

$$P_y = p_c + \frac{P(\delta_o + \delta)h}{Ary^2} \quad ...(8)$$

$$p_y = p_o + p_c \frac{(\delta_o + \delta)h}{ry^2} \quad ...(9)$$

$$p_y = p_c + \frac{p_c h}{ry^2}\left[\delta_o\left(1 + \frac{1}{(p_E\, p_c) - 1}\right)\right] \quad ...(10)$$

Denoting the Perry factor,

$$\eta = \frac{\delta_o h}{ry^2} \quad ...(11)$$

$$(p_y - p_c) = \eta p_c \left[1 + \frac{p_c}{(p_E p_c)}\right] \quad ...(12)$$

On simplification it gave,

$$(p_E - p_c)(p_y - p_c) = \eta p_E\, p_c \quad ...(13)$$

The value of pc, the limiting strength at which the maximum stress equal the design strength, can be found by solving this equation and η is the Perry factor.

3.1.2 Basis of Current Codal Provision for Compression Member Design

The design compressive stress, f_{cd}, of axially loaded compression members shall be calculated using the following equation,

$$f_{cd} = \frac{f_y / \gamma_{mo}}{\phi + \left[\phi^2 - \lambda^2\right]^{0.5}} = \chi f_y / \gamma_{mo} \leq f_y / \gamma_{mo}$$

Where,

$$\phi = 0.5\left[1 + \alpha(\lambda - 0.2) + \lambda^2\right]$$

$\lambda =$ non- dimensional effective slenderness ratio.

$$= \sqrt{f_y / f_{cc}} = \sqrt{f_y \left(KL/r\right)^2 / \pi^2 E}$$

$$f_{cc} = \text{Euler buckling stress} = \frac{\pi^2 E}{\left(KL/r\right)^2}$$

Where,

$KL / r =$ effective slenderness ratio or ratio of effective length, KL to appropriate radius of gyration r.

$\alpha =$ imperfection factor given in table.

$\chi =$ stress reduction factor for different bucking class. Slenderness ratio and yield stress.

$$= \frac{1}{\left[\phi + \left(\phi^2 - \lambda^2\right)^{0.5}\right]}$$

$\lambda_{mo} =$ partial safety factor for material strength.

Effective Length of Compression Members

The effective length KL is calculated from the actual length L of the member considering the rotational and relative translational boundary conditions at the ends. The actual length shall be taken as the length from centre-to-centre of its intersections with the supporting members in the plane of the buckling deformation. In the case of a member with a free end the free standing length from the center of the intersecting member at the supported end shall be taken as the actual length.

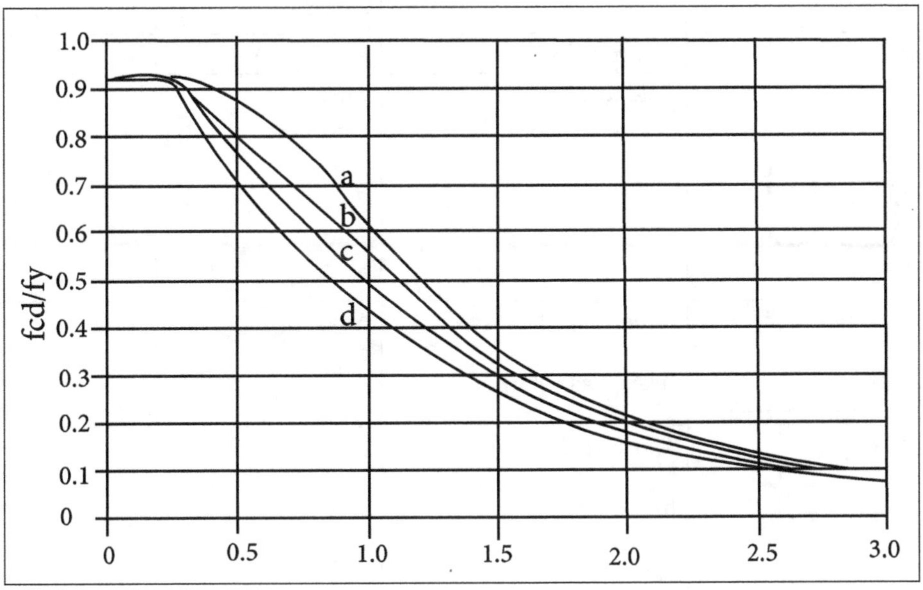

Column buckling curves.

Table: Buckling Class of Cross – Section:

Cross- Section (1)	Limits (2)	Buckling About Axis (3)	Buckling Class (4)
Rolled I-Section 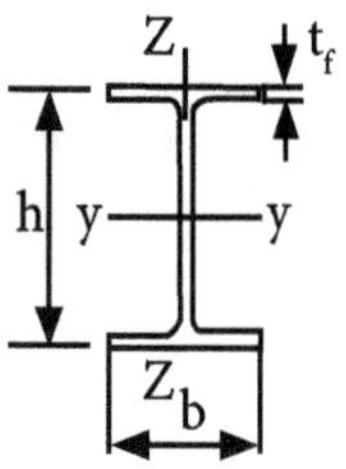	h/b >1,2 $t_f \le 40$ mm $40 \text{ mm} < t_f \le 100\text{mm}$	y – y z – z y – y z – z	a b b c
	h/bt ≤ 1, 2 : tf ≤ 100 mm tf > 100 mm	z – z y – y z – z y - y	b c d d
Welded I-Section 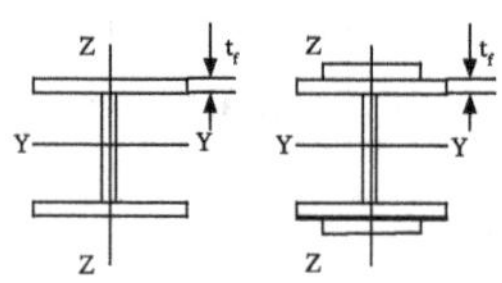	tf ≤ 40 mm tf > 40 mm	z – z y – y z – z y – y	b c c d
Hollow Section 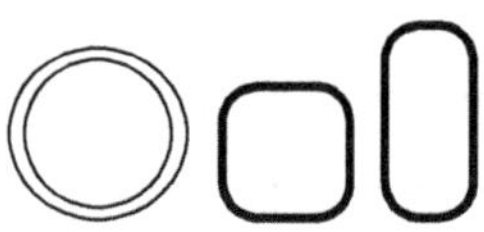	Hot rolled	Any	a
	Cold formed	Any	b
Welded Box Section 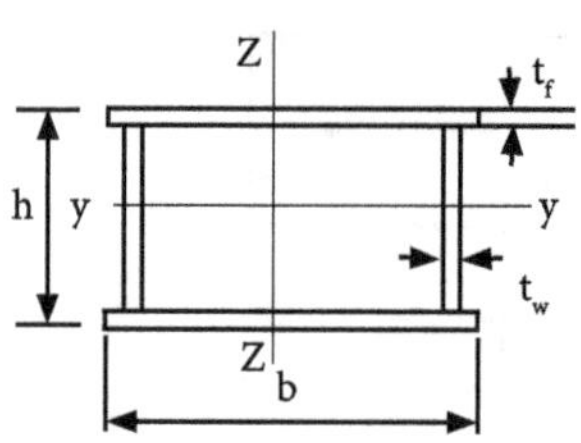	Generally (except as below)	Any	b
	Thick welds and h/tf < 30 h/tf < 30	z – z y - y	c c

Channel, Angle, T and Solid Sections 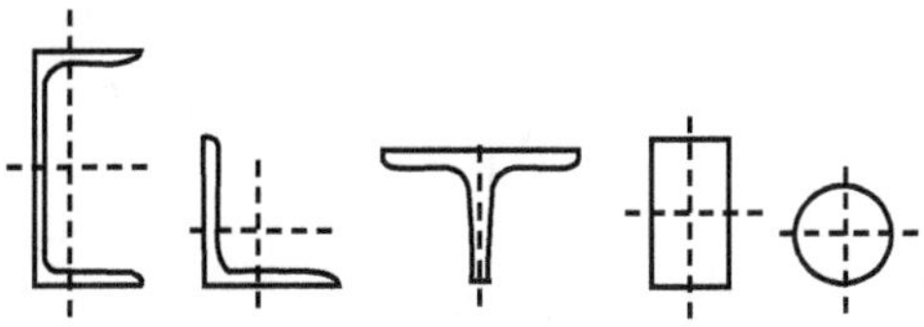	Any	c
Built-up Member 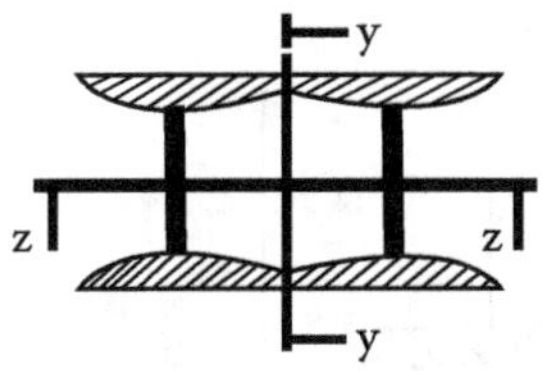	Any	c

Table: Effective Length of Prismatic Compression Members:

Boundary Conditions				Schematic Representation	Effective Length
At One End		At the Other End			
Translation	Rotation	Translation	Rotation		
(1)	(2)	(3)	(4)	(5)	(6)
Restrained	Restrained	Free	Free		2.0 L
Free	Restrained	Free	Restrained	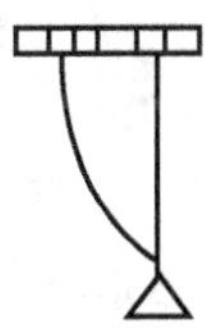	
Restrained	Free	Restrained	Free	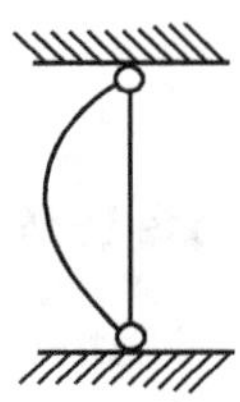	1.0 L

Restrained	Restrained	Free	Restrained		1.2 L
Restrained	Restrained	Restrained	Free	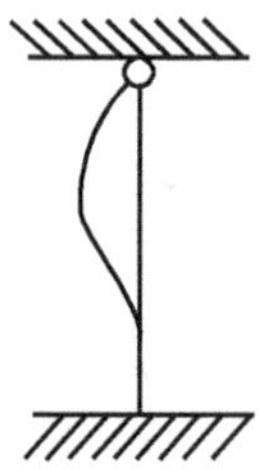	0.8 L
Restrained	Restrained	Restrained	Restrained		0.65 L

3.1.3 Slenderness Ratio

The slenderness ratio is defined as the ratio of effective length or height of a structural member to its least radius of gyration,

$$\text{Slenderness ratio} \quad \frac{\text{Effective height}}{\text{Least radius of gyration}}$$

3.2 Design of Single Section and Compound Section Compression Members

1. Let us discuss a column to carry a factored axial compressive toad of 500 kN. The effective length of the column is 4m. fy = 250 MPa.

Solution:

Let us choose ISMB 300.

Column → Pinned@both ends.

f_y = 250 MPa.

1. Properties of Section

f_y = 250 MPaf_{ck} = 410MPa

A_e = A = 5626 nun²E = 2 × lo⁵MPa.

Effective Length [KL] for both ends pinned,

h = 300 mm.

b_f = 140 mm.

t_f = 12.4 mm.

γmin = γ_y = 28.4mm.

2. Buckling Class Determined

[As per T able 10 of IS 8001],

$$\frac{h}{b_f} = \frac{3000}{140} = 2.14$$

$$t_f = 12.4 \text{ [i.e.]} \leq 100 \text{ mm.}$$

The classification for buckling class is a falls. It falls under buckling Class 'b' for bucking about ZZ-axis and buckling Class 'c' for about YY-axis.

3. Euler Buckling Stress, f_{cc}

$$f_{cc} = \frac{\pi^2 E}{\left[\frac{KL}{\gamma_{min}}\right]^2} = \frac{\pi^2\left[2\times10^5\right]}{\left[\frac{4000}{28.4}\right]^2}$$

$$= 99.5 \text{ MPa.}$$

4. Calculation of Effective Slenderness Ratio [λ]

[as PerCl.7.12.1 of IS800],

$$\lambda = \sqrt{\frac{f_y}{f_{cc}}}$$

$$= \sqrt{\frac{250}{99.5}}$$

$$\lambda = 1.585$$

5. Computing Off$_{cd}$

Ag per cl.7.1.2.1 of I$_s$ 800:2007,

$$fcd = \frac{f_y / \gamma_{mo}}{\phi + \left[\phi^2 - \lambda^2\right]^{0.5}} \le \frac{f_y}{\gamma_{mo}}$$

$$\phi = 0.5\left[1 + \alpha\left[\lambda - 0.2\right] + \lambda 2\right]$$

α = 0.34 for buckling class 'b' about

$$\phi = 0.5\left[1 + 0.34\right]\left[\lambda - 0.2\right] + \lambda_2$$

$$\phi = 1.36$$

γ_{mo} = 1.1 [as per Table IS 800:2007]

$$f_{cd} = \frac{250 / 1.1}{1.36 + \left[1.36^2 - 1.18^2\right]^{0.5}}$$

$$f_{cd} = 111.62 \le 250/1.1.$$

6. Computation of Strength of Column

$$P_d = A_g f_{cd}$$

$$= 5626 \times 111.62$$

$$= 629794 \text{ N}$$

$$P_d = 627.9 \text{ KN} > 500 \text{ KN}.$$

Hence chosen section is safe.

2. Here we solve a built-up column with two channel sections. The column is of 6.4 m effective length and supports a load of 1000kN.

Solution:

Assuming a safe stress of loo N/mm².

$$\text{Approximate area required} = \frac{1000 \times 1000}{100} \text{ mm}^2$$

$$= 10{,}000 \text{ mm}^2.$$

Let us try two channels ISLC 350, 38.8 kg/m.

$$a = 4.947 \times 2 = 9894 \text{ mm}^2$$

$$\gamma_{x2} = 137.2 \text{ mm}.$$

γ_{yy} will be made equal to γ_{xx} by choosing proper spacing between the channels.

$$\text{Slenderness ratio} = \frac{l}{\gamma} = \frac{10 \times 1000}{137.2}$$

$$= 72.9.$$

Safe compressive stress corresponding to the above slenderness ratio = 108.8 N/mm²,

$$\text{Induced stress} = \frac{1000 \times 1000}{9894} = 101.1 \text{ N/mm}^2.$$

∴ The design is safe.

Let the spacing between the channels be 'b',

Approximately,

$$\gamma_{yy} = 0.6 \text{ b}$$

But,

$$\gamma_{xx} = 137.2 \text{ mm}$$

$$\gamma_{yy} = \gamma_{xx}$$

$$0.6b = 137.2$$

$$b = \frac{137.2}{0.6} = 228.7 \text{ mm.}$$

Provide a spacing 6 = 230 mm between the channels, now the column section.

$$I_{yy} = 2 \times 3.946 \times 10^6 + 2{,}4947 \, [115 + 24.1^2]$$

$$= 199.33 \times 10^6 \text{ mm}^4$$

$$\gamma_{yy} = \sqrt{\frac{199.33 \times 10^6}{9{,}894}} = 141.9 \text{ mm}$$

$$\gamma_{yy} = 137.2 \text{ mm}$$

$$\gamma_{xz} < \gamma_{yy}.$$

Hence slenderness ratio adopted earlier holds good.

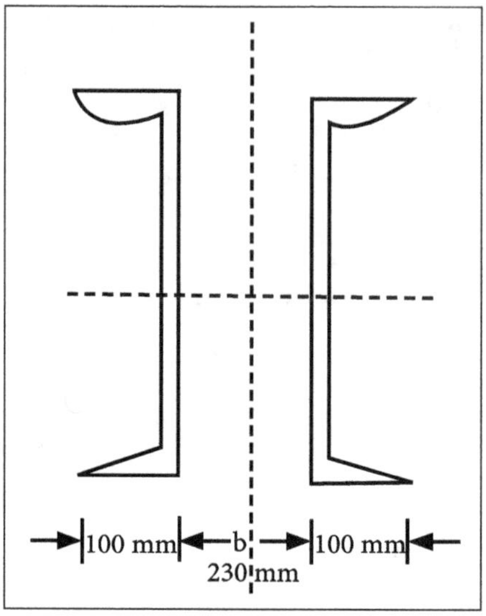

3.3 Design of Laced and Battened Type Columns

Design of Laced Columns

As far as possible the latticing system shall be uniform throughout:

- In single laced system the direction of lattices or opposite faces should be shadow of the other. It should not be mutually opposite.
- In bolted/riveted construction, the minimum width of lacing bars shall be three times the nominal diameter of the bolts/rivets.
- The thickness of flat lacing bars shall not be less than 1/40th of its effective length for single lacing and 1/16th of the effective length for double lacings.
- Lacing bars shall be inclined at 40° to 70° to the axis of built up member.
- The distance b/w the two main members should be kept so as to get ryy > rzz where ryy is the R.o.G about weaker axis and rzz is the R.o.G of stronger axis of individual member.
- Maximum spacing of lacing bars shall be such that the maximum slenderness of the main member b/w consecutive lacing connection is not greater than 50 or 0.7 times the most unfavorable slenderness ratio of the member as a whole.
- The lacing shall be designed to resist transverse shear vt = 2.5% of axial force in column. If there are two transverse parallel system then each system has to resist vt/2 shear force.

- If the column is subjected to bending also, vt = bending shear + 2.5% column force.
- Effective length of single laced system is equal to the length between the inner end fastener. For welded joints and double laced effectively connected t intersection effective length may be taken as 0.7 times the actual length.

Design of Batten Columns

Batten plates should be provided symmetrically:

- At both ends batten plates should be provided. They should be provided at point where the member is stayed in its length.
- The number of battens should be such that the member is divided into not less than three bays as far as they should be spaced and proportioned uniformly throughout.
- Battens shall be of plates, angles, channels, or I-sections and at their ends shall be riveted, bolted or welded.
- By providing battens distance between the member of columns is co maintained that r yy >r xx.
- The effective slenderness ratio of battened columns shall be taken as 1.1 times the maximum actual slenderness ratio of the column to account for shear deformation.
- The vertical spacing of battens, measured as centre to centre of its end fastening shall be such that the slenderness ratio of any component of column over that distance shall be neither greater than 50 nor greater than 0.7 times the slenderness ratio of the member as a whole above its z-z axis.
- Battens shall be designed to carry the bending moments and shear force arising from transverse shear force vt equal to 2.5% of the total axial load.
- In case columns are subjected to moments also the resulting shear force should be found and then the design shear is sum of this shear and 2.5% of axial load.
- In case columns are subjected to moments also the resulting shear force should be found and then the design shear is sum of this shear and 2.5% of axial load.
- The design shear end moments for battens plates is given by Vb = Vt x C/NS and M = Vt x C/ 2N at each connection.

V_t = transverse shear force.

C = distance b/w Centre to Centre of battens longitudinally.

N= number of parallel planes.

S = minimum transverse distance b/w the centroid of the fasteners connecting batten to the main member.

The effective depth of end battens (longitudinally), shall not be less than the distance b/w the centroids of main members.

Effective depth of intermediate battens shall not be less than ¾ th of the above distance.

In no case the width of battens shall be less than twice the width of one member in the plane of the batten. It is to be noted that the effective depth of the batten shall be taken as the longitudinal distance b/w the outermost fasteners.

The thickness of battens shall be not less than 1/50th of the distance between the innermost connecting lines of rivets, bolts or welds.

The length of the weld connecting batten plate to the member shall not be less than half the depth of batten plate. At least one third of the weld shall be placed at each end of this edge.

Problems

1. Let us discuss a laced column with two channels back to back of length 10 m to carry an axial factored load of 1400 kN. The column may be assumed to have restrained in position but not in direction at both ends (hinged ends).

Solution:

$$\text{Assuming } fcd = 135\ N/mm^2$$

$$\text{Area required} = \frac{1400 \times 1000}{135} = 10370\ mm^2$$

Try 2 ISMC 350@ 413 N/m.

$$\text{Area provided} = 2 \times 5366 = 10732\ mm^2$$

Distance will be maintained so as to get $r_{yy} > r_{zz}$.

$$\therefore \text{Actual} = \frac{KL}{r} = \frac{1 \times 10000}{136.6} = 73.206.$$

Since it is a laced column,

$$\frac{KL}{r} = 1.05 \times 73.206 = 76.87.$$

From Table,

$$f_{cd} = 152 - \frac{6.87}{10}(152 - 136)$$

$$= 141.0 \text{ N/mm}^2.$$

Load carrying capacity,

$$= 10732 \times 141.0$$

$$= 1513.297 \times 10^3$$

$$= 1513.297 \text{ kN} = 1400 \text{ kN O.K.}$$

Spacing between the channels.

Let it be a clear distance "d".

Now,

$$I_{xx} = 2 \times 10008 \times 10^4 = 20016 \times 10^4 \text{ mm}^4.$$

$$I_{yy} = 2\left[430.6 \times 10^4 + 5366\left(\frac{d}{2} + 24.4\right)^2\right]$$

Equating I_{yy} to I_{xx},

We get,

$$2\left[430.6 \times 10^4 + 5366\left(\frac{6}{2} + 24.4\right)^2\right] = 20016 \times 10^4$$

i.e.,

$$\left(\frac{d}{2} + 24.4\right)^2 = 17848.3$$

Provide d = 220 mm as shown in Figure.

Let the lacings be provided at 45° to horizontal.

Horizontal spacing of lacing,

= 220 + 60 + 60

= 340 mm [Note: g = 60 is gauge distance].

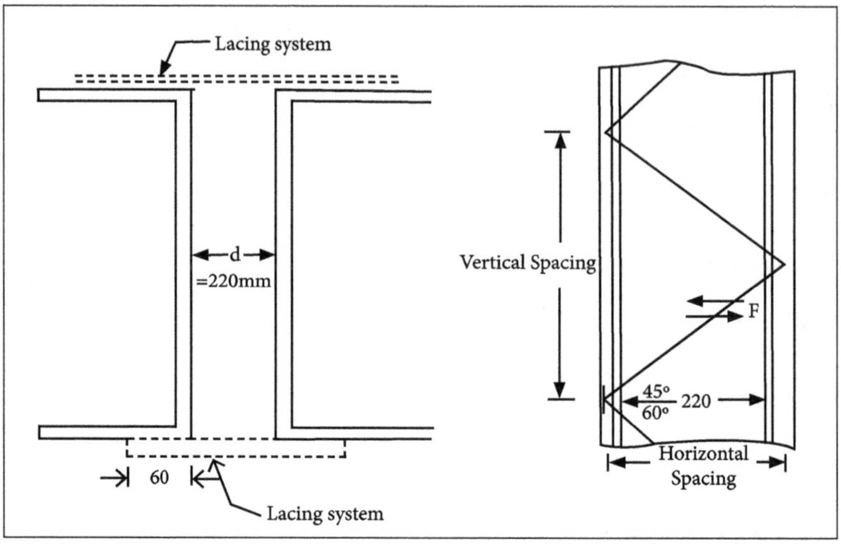

Least r of one channel = r_{yy} = 28.3

Transverse shear to be resisted by lacing systems,

$$=\frac{2.5}{100}\times 1400 \times 10^3 = 35000 \text{ N.}$$

Shear to be resisted by each lacing systems,

$$=\frac{35000}{2}=17500 \text{ N.}$$

Length of lacing $=(220 + 60 + 60)\ \frac{1}{\cos 45} = 480.83 \text{mm.}$

Minimum thickness of lacing $=\frac{1}{40}\times 480.83$

$$= 12.02 \text{ mm.}$$

Sectional area = 60 × 14 = 840 mm²,

$$r_{min}=\frac{\sqrt{\frac{1}{12}\times 60\times 14^3}}{60\times 14}=4.041 \text{ mm.}$$

$$\frac{KL}{r}=\frac{480.3}{4.641}=18.97 \times 14.5$$

Strength of 20 mm Shop Bolts:

(i) In single shear,

$$=0.78 \times \frac{\pi}{4}\times 20^2 \times \frac{400}{\sqrt{3}\times 1.25}=45272 \text{ N}$$

Edge distance $=\frac{60}{2}=30.$

$$\therefore \text{ Kb} = \frac{30}{3\times 22}=0.4545$$

(ii) Strength in bearing $= \frac{2.5\ K_b\ dt\ f_u}{1.25}$

$$= \frac{2.5 \times 0.4545 \times 20 \times 10 \times 400}{1.25}$$

$$= 101808 \text{ N.}$$

Bolt value = 45272.

Number of bolt required $= \frac{17500}{45272}=0.387$

Provide one bolt.

Check for the Strength of Lacing:

$$\frac{KL}{r}=118.97$$

A flat belongs to bucking class c,

$$f_{cd} = 94.6 - \frac{8.97}{10}(94.6 - 83.7)$$

$$= 84.82 \text{ N/mm}^2.$$

Load carrying capacity in compression,

$$= 84.82 \times 60 \times 14$$

$$= 71251 \text{ N.}$$

2. Let us discuss a laced column for an axial load of 1200 kN with an effective span of 7.5 m has one end fixed out the other end hinged. Use channels for main members and an angle for lacing bars.

Solution:

Step 1: Let us assume f_{cd} = 135 MPa.

Step 2: Gross area required for the column $= \frac{P_d}{f_{cd}}$.

Design load for the column,

$= (1.5)(1200)$

$= 1800$ kN

Gross area required,

$$= \frac{1800 \times 1000}{135}$$

$= 13333.33$ mm².

Step 3: Let us assume that the channels are blacked back to back.

Let us try with 2 channels 23 MC 400 at 484.5 N/N.

- Giving an area of 2(6293) = 12586 mm2.
- γxx of one 1 SMC 400 = 154.8 mm.
- Least radius of gyration γxx = 154.8 mm.
- The effective length KL is given as = 7500 mm.

$$\frac{KL}{\gamma} = \frac{7500}{*} = *$$

Design the two channels as connected by Larcey's,

$$\left(\frac{KL}{\gamma}\right)_e = 1.05\left(\frac{KL}{\gamma}\right)_o$$

$$\left(\frac{KL}{\gamma}\right)_e = 1.05\,(41.34)$$

$$= 43.407$$

Step 4: Refer table of IS 800,

$$\frac{KL}{\gamma} = 43.407$$

$f_y = 250$ MPa

$$f_{cd} = 198\text{-}\left(\frac{198-183}{10}3.407\right)$$

$$= 192.89 \text{ MPa.}$$

Load carrying capacity,

$$= f_{cd}\ (A)$$

= 192.89 (12586)

= 2427714 N

= 2427.714 kN

Step 5: Spacing between the channels.

From steel table for one 1 SMC,

$$I_{XX} = 15082.8 \times 10^4 \text{ mm}^2$$

For two channels,

$$I_{xx} = 2\left(15082.8 \times 10^4\right) \text{mm}^4$$

$$I_{YY} = 2\left[504.8 \times 10^4 + 6293\left(\frac{d}{2} + 29.2\right)\right]$$

$$I_{XX} = I_{XX}$$

$$d = 256 \text{ mm.}$$

Let us provide 260 mm as the distance between 2 channels.

Step 6: Design of lacings.

Let the lacings be provided at 45° to the horizontal and let the gauge distance for the bolt be 60 mm.

Horizontal spacing of lacings,

= 260 + 60 + 60

= 380 mm.

The vertical spacing (1),

$= 2\ (380\ \tan 45^{\circ})$

$= 760$ mm.

Least value of γ_{min} = 28.3 mm for ISMC 4 N,

$$\frac{a_1}{n} > 50$$

$$\frac{a_1}{n} = \frac{760}{28.3} = 26.855 < 50.$$

Hence OK.

Transverse shear to be resisted by lacing systems = 2.5%.

Axial force in the column,

$$= \frac{2.5}{100}\left(1800 \times 10^{3}\right)$$

$= 45000$ N.

Shear to be resisted by each lacing,

$$= \frac{45000}{2} = 22500 \text{ N.}$$

Length of Lacing,

$$= (260 + 60 + 60)\frac{1}{\cos 45^{\circ}}$$

$= 537.40$ mm.

Minimum width of lacing as per d.7.6.2 of IS 800°C 2007,

= 3 × (nominal diameter of end bolt)

= 3 × 20

= 60 mm (assuming 20 mm ϕ bolt)

Minimum thickness of lacing as per Cl. 7.6.3 of IS 800°C 200),

$$= \frac{1}{40} \text{ (Effective length of lacing)}$$

$$= \frac{1}{40}(537.40)$$

$= 13.435$ mm.

Let us provide 14 mm thick flats for the lacing.

Hence let us use 60 ISF 14 for lacing.

Sectional area,

$$= 60 \times 14$$

$$= 840 \text{ mm}^2.$$

$$\gamma_{min} = \sqrt{I/A}$$

$$= \sqrt{\frac{60 \times 14^3/12}{60 \times 14}}$$

$$= 4.041 \text{ mm}$$

$$\frac{KL}{\gamma} = \frac{L}{r} = \frac{537.40}{4.041} = 132.99 < 145.$$

Hence OK.

Step 7: Design of bolt for the lacings.

Assuming 20 mm bolts

Strength of one bolt in single shear,

$$= 0.78\left(\frac{\pi(20)^2}{4}\right)\frac{400/\sqrt{3}}{1.25}$$

$$= 45272 \text{ mm.}$$

Edge distance $= \frac{60}{2} = 30 \text{ mm}$

$$k_b = \frac{P}{3d_o} = \frac{30}{3(22)} = 0.4545$$

Strength of one bolt in bearing,

$$= \frac{2.5\, k_b\, dt\, f_u}{1.25}$$

$$= \frac{2.5\,(0.4545)(20)(14)(410)}{1.25}$$

$$= 104353 \text{ N.}$$

Bolt Value = 45272 N.

No. of bolts $= \frac{18750}{45272} = 0.0414$

Hence let us provide one bolt for the connection at each end of the lacing.

Step 8: Check for strength of lacing,

$$\frac{KL}{\gamma} = 132.99$$

Referring to Table of IS 800: 2007,

$$f_{cd} = 74.3 - \frac{74.3 - 66.2}{10}$$

$$= 2.99$$

$$f_{cd} = 71.88 \text{ MPa.}$$

Load carrying capacity in compression of lacing,

$$= f_{cd}$$

$$= 71.88\,(60 \times 14)$$

$$= 60379 \text{ N.}$$

3. Let us discuss a column with single lacing system to carry a factored axial load of 1500 kN. The effective net of the column is 4.2 m. Use two channels placed toe to toe.

Solution:

Step 1: Let us assume the value of f_{cd} = 135 MPa.

Step 2: Gross area assume for the column $A_g = \frac{P_d}{f_{cd}}$. Axial load P_d = 1500 kN.

$\therefore$ Gross area required $= \frac{1500(1000)}{135}$

$= 1111 \text{ mm}^2.$

Step 3: γ_{xx} of one IS MC = 154.8 mm

Least radius of gyration = γ_{xx} = 154.8 mm

$\therefore$ The effective length,

$$\frac{K_L}{r} = \frac{4200}{154.8} = 27.13$$

IS 800 – 2007,

$$\left(\frac{K_L}{r}\right)_2 = 1.05\left(\frac{K_L}{r}\right)_0 \text{ where}$$

$$\left(\frac{K_L}{r}\right)_e = 1.05\ (27.13)$$

$$= 28.487.$$

Step 4: $f_{cd} = 224 - \left(\frac{224-211}{10}\right)(8.487)$

$$= 212.97 \text{ MPa.}$$

Load carrying capacity,

$= f_{cd}$ (A)

= 212.97 (12586)

= 2680440 N

= 2680.44 kN > 1500 kN.

Step 5: $I_{XX} = (15082.8 \times 10^4)$ mm^4, $C_{XX} = 24.2$ mm

$$I_{XX} = 2(15082.8 \times 10^4) \text{ mm}^4$$

$$I_{yy} = 2\left[504.8 \times 10^4 + 6293\left(\frac{d}{2} - 24.2\right)^2\right]$$

$$= 2(15082.8 \times 10^4)$$

$$d = 352.8 \text{ mm.}$$

Step 6: Design of lacings

∴ Horizontal spacing of lacings = 360 + 60 + 60 = 480 mm

∴ The vertical spacing of lacing bar,

= 2(360 – 60 – 60) (tan 45°)

$= 2(240 \tan 45°) = 480$ mm.

$$= \frac{a_1}{r_1} = \frac{480}{28.3} = 16.96 < 50$$

Transfer shear system = 2.5%, axial force column,

$$= \frac{2.5}{100}(1500 \times 10\ ) = 37500 \text{ N}$$

∴ Shear resisted = $\frac{37500}{2}$

= 18750 N.

Minimum width of lacing:

= 3 × (nominal diameter)

= 3 × 20 = 60 mm

Minimum Thickness,

$$= \frac{1}{40} \text{ (effect length)}$$

$$= (360 - 50 - 50)\ \frac{1}{\cos 45°}$$

= 367.7 mm.

$$\gamma_{min} = \sqrt{I/A}$$

$$= \sqrt{\frac{60 \times 10^3/12}{60 \times 10}} = \frac{t}{\sqrt{12}} = \frac{10}{\sqrt{12}}$$

$$\frac{K_L}{r} = \frac{l}{r} = \frac{367.7}{2.89} = 127.33 < 145.$$

Step 7: Design of bolt for the lacing,

$$= 0.78\left(\frac{\pi(20)^2}{4}\right)\frac{400/\sqrt{3}}{1.25}$$

$$= 45272 \text{ N}$$

$$= \frac{60}{2} = 30\text{mm}$$

$$k_b = \frac{P}{3d_o} = \frac{3}{3(22)} = 0.4545.$$

Strength of one both bearing,

$$= \frac{2.5\ k_b\ dt\ fy}{1.25}$$

$$= \frac{2.5\ (0.4545)(20)(10)(410)}{1.25}$$

$$= 45272\ \text{N}$$

No. of bolt = $\frac{18750}{45272} = 0.414$

$$\frac{k_L}{r} = 127.23$$

Check for strength of lacing,

$$f_{cd} = \frac{88.7(83.7 - 7413)(-1.23)}{10}$$

$$= 76.90\ \text{MPa.}$$

Load carrying capacity = f_{cd} (area of cross section),

= 76.90 (60 × 10) = 46142 N

Force in lacing = $\frac{18750}{\sin 45°}$ = 26516 N < 46142 N.

∴ Let us provide 60 IS flat is Inclination and connect them to centre of gravity of channels with one bolt of 20 mm nominal diameter at each end.

3.4 Design of Column Bases

Steps in the Design of a Column Slab Base Connection

- Maximum bearing pressure of concrete for the slab base * 0.6fck.
- Area of base plate.

- Thickness of base plate.
- Connection.

Problems

1. A column of ISMB 400 is subjected to an axial force of 750 kN. Let us design suitable base plate and assume necessary data's required.

Solution:

Design:

Step 1: IS 800 : 2007

Bearing strength of concrete pedestal,

$= 0.45\ f_{cl}$

$= 0.45\ (20)$

$= 9$ MPa.

Assuming a load factor 1.5

Factored load $p_u = 1.5\ (750)$

$= 1125$ kN.

Step 2: Area of base plate.

Area of base plate required,

$$= \frac{1125 \times 1000}{9}$$

$= 125000\ mm^2$.

The properties of IS MB 400,

D = 400 mm.

B = 140 mm.

$t_f = 16$ m.

$t_w = 8.9$ m.

Provide an equal projection of summon all the 4 edges of the column.

Size of base blase provided = $[400 + 2(60)] \times (140 + 2(f_o))$

$\Rightarrow 520\ mm \times 260\ mm$

Area of base provided = 135200 mm^2,

= 125000 mm^2.

Step 3: Thickness of base plate.

Actual pressure at base $W = \dfrac{7125 \times 10^3}{520 \times 260}$

W = 8.32 MPa,

$$t_s = \sqrt{2.5\, w\left(a^2 - 0.3\, b^2\right)\gamma_{mo}\; f_y} > t_f$$

W = 8.32 MPa, a = b = 60 mm, γ_{mw} = 1.1, f_y = 250 MPa

t_f = 16 mm for given ISMB 4 av

$$t_s = \sqrt{2.5\,(7.32)\left(60^2 - 0.3(60)^2\right)(1.1)/250}$$

= .15.18 > 16 mm

= 15.18 mm should be t_7.

Hence let us bromide a thickness of 16 mm for base plates.

Step 4: Connection.

Let us use 4 bolts of 20 mm diameter and 300 mm long to anchor the blade of size,

520 × 260 × 16 mm to the concrete formulation.

Welds: The length available for welding

= 2 [140 + (140 – 16) + 400 – 2(7.5)]

= 1298 mm

Strength of weld/mm $= \dfrac{410\sqrt{3}}{1.25}$

= 189.3 MPa.

Effective area of weld = 0.7 (t) (L_e)

0.7 (t) (L_e) (189.38) = 1125 × 10^3

$$t = \frac{1125 \times 10^3}{0.7 \times 189.38 \times 1298}$$

t = 6.53 mm.

Hence let us provide 8 mm as the size of weld.

2. Let us discuss a steel column of rolled steel I section to carry an axial load of 500 kN. The column is 4 m long and it is effectively held in position at both ends but restrained against at one end only. Take yield stress in steel as 250 N/mm².

Solution:

Design a Steel Column:

- Rolled Steel I Section.
- Arial load, P_d = 500 kN.
- Column = 4 m long.

Effectively held @ both ends in position.

One end is restrained,

f_y = 250 N/mm2.

⇒ Selection of Suitable Section:

$$\left.\begin{array}{c}\text{One end restrained}\\ \text{Other end held in position}\end{array}\right\} \Rightarrow \text{KL Effective Length} = 1.2\text{ L}$$

Refer as per Table – IS 800:2007.

Effective Length,

= 1.2L = 1.2 × 4000

= 6000 mm.

Assume design stress for an of section f_{cd} = 150Mpa,

Gross Section Area required,

$$A_g = A = \frac{P_d}{f_{cd}} = \frac{500000}{150} = 3333 \text{ mm}^2$$

By referring steel table choosing, ISMB 225,

A_g = 39723 mm².

h = 225 mm.

b_f = 110 mm.

L_c = 11.9 mm.

t_w = 6.5 mm.

r_{YY} = 23.4 mm.

Buckling Class:

As per Table of IS800-2007.

$$\frac{h}{b_f}=\frac{225}{250}=0.9 \ \& \ t_f=11.8mm \le 40 \text{ mm.}$$

$$\frac{h}{b_f}\le 1.2 \ t_f \ \le 100 \text{ mm.}$$

It falls under buckling Class 'b' for buckling about ZZ-axis and buckling Class 'c' for buckling about YY-axis.

Computation of Euler Buckling Stress:

$$f_{cc}=\frac{\pi^2 E}{\left[\frac{KL}{r_{min}}\right]^2}=\frac{\pi^2\times 2\times 10^5}{\left(\frac{4000}{23.4}\right)^2}=67.55/\text{mm}^2$$

Calculation of Euler Slenderness Ratio, λ:

As per CI.7.1.2.1 of IS800-2007,

$$\lambda=\sqrt{\left(\frac{f_y}{f_{cc}}\right)}=\sqrt{\frac{250}{67.5}}=1.92$$

Computation off_{cd}:

As per CI.7.1.2.1 of IS800-2007,

$$f_{cd}=\frac{f_y/\gamma_{m_o}}{\phi+\left[\phi^2-\lambda^2\right]0.5}\le\frac{f_y}{\gamma_{m_o}}.$$

$$\phi=0.5\left[1+2(\lambda-0.2)+\lambda^2\right]$$

$$\alpha=0.34\ [\text{Table of IS800}-2007]$$

$$\phi=0.5\left[1+0.34[1.92-0.2]+1.92^2\right]$$

$$=\ 2.6356$$

$$f_y=250\text{ Mpa}$$

$$\gamma_{m_o=}\ 1.1\ [\text{Tableof IS800}-2007]$$

$$f_{cd} = \frac{250/1.1}{2.6356 + \left[2.6356^2 - 1.92^2\right]^{0.5}} < \frac{250}{1.1}$$

$$= 51.17\ 227.27\ \text{MPa.}$$

Computation of Strength of Column:

$$P_d = A_g f_{cd}$$

$$= 3972 \times 51.17$$

$$= 203247\ \text{N}\ 500000\text{N.}$$

3. Let us discuss a buildup column composed of two channel sections placed back to back, carrying an axial load of 1345 kN. Effective length of column is 5.95 m. Take fy = 250 kN/mm².

Solution:

Design a Buildup Column,

Two Channel Section – back to back,

Arial Load = 1345 kN.

Effective length of column = 5.95 m.

f_y = 250 kN/mm².

Assume f_{cd} = 135 MPa.

Gross Section Area Required:

$$A_g = \frac{P_d}{f_{cd}}$$

$$= \frac{1345 \times 1000}{135} = 9963\ \text{mm}^2$$

Let us try with 2 channel ISMC 400 @ 484.5 N/m.

Giving an area of 2[6293] = 12586 mm².

r_{xx} of one ISMC 400 = 154.8 mm.

Spacing between the channels will be maintained such that $\gamma_{yy} \geq Y_{xx}$, so that.

Least rating of gyration γ_{xx} = 154.8 mm.

Effective Length = KL = 5950 mm.

$$\frac{KL}{r}=\frac{5950}{154.8}=38.4$$

Design the two channels, as connected by Lacing.

As per CI.7.6.1.5 of IS800-2007,

$$\left(\frac{KL}{r}\right)_e=1.05\left[\frac{KL}{r}\right]_o$$

$$=1.05[38.4]=40.35$$

*Refering to Table of IS800-2007, built up Sections belong to buckling Class 'C',

$$\frac{KL}{r}=38.4\ f_y=250\ \text{Mpa},$$

From Table C of IS800-2007,

$$f_{cd} = 201\ \text{MPa}.$$

Load carrying capacity,

$$= f_{cd} \times \text{Area}$$

$$= 201 \times 12586$$

$$= 2427714\ \text{N}\ 1345000\text{N}.$$

Spacing between the Channels:

Let 'd' be the distance between the channel.

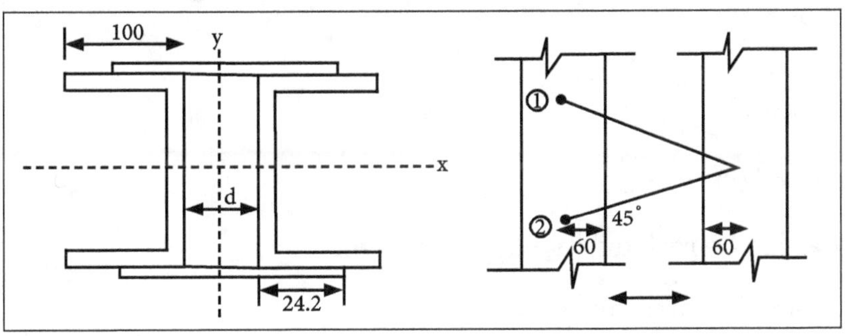

From Steel Table per one ISMC 400,

$$I_{xx}=15082.8\times 10^4\ \text{mm}^4$$

$$I_{xx}=2\left[504.8\times 10^4 + 6293\,[d/2 + 24.2]^2\right]$$

$$I_{YY} = I_{XX}$$

$$2\left[504\times10^4 + 6293\left(\frac{d}{2}\times24.2\right)^2\right] = 2\times15082.8\times10^4$$

$$d = 256 \text{ mm} \approx 260 \text{ mm.}$$

Design of Lacings:

Let the Lacings be provide @ 45° to the horizontal and let the gauge distance for the bolt 60 mm.

Horizontal Spacing of Lacing,

= 260 + 60 + 60

= 380 mm.

Vertical Spacing (1), (2),

$= 2 \times 380 \times \tan 45°$

= 760 mm.

$\gamma_{min} = 28.3$ mm.

Referring to CI.7.6.5.1,

$$\frac{a_1}{r_1} > 50$$

$$\frac{760}{28.3} > 50$$

$$26.85 > 50.$$

As per CI.7 6.6 1 of IS800-2007.

Transverse Shear to be resisted by Lacing System,

= 2.5% of axial forces in column

$$= \frac{2.5}{100} \times 1345\times10^3 = 33625 \text{ N.}$$

∴ Shear to be resisted by each Lacing = 16812.5 N.

Length of Lacing,

$$= [260 + 60 + 60] \frac{1}{\cos 45^\circ}$$

$$= 537.4 \text{ mm.}$$

As per CI.7.6.2 of IS800-2007,

Minimum width of Lacing,

$$= 3 \times d_n$$

$$= 3 \times 20 = 60 \text{ mm}$$

[assume 20 mm and bolt]

As per CI.7.6.3 of IS800-2007.

Minimum thickness of lacing = $3 \times d_n$

[assume 20 mm and bolt]

As per CI.7.6.3 of IS800-2007.

Minimum Thickness of Lacing:

$$= \frac{1}{40} \text{ Effective length of Lacing}$$

$$= \frac{1}{40} \times 537.4$$

$$= 13.435 \approx 14 \text{ mm}$$

Sectional Area = 60 × 14 = 840 mm²,

$$\gamma_{min} = \sqrt{I/A} = \sqrt{\frac{60 \times 14^3}{12 \times 14}} = 4 \text{ mm}$$

By referring to CI.7.6.6.3 of IS800-2007.

$$KL = L$$

$$\frac{L}{r} < 145$$

$$\frac{537.40}{4.041} = 132.99 < 145$$

Hence Okay.

Design of Bolt for the Lacings:

Assume ϕ 20 mm bolt.

Strength of one bolt in single shear,

$$=0.78\left[\frac{\pi}{4}20^2\right]\frac{400}{\sqrt{3}\times 1.25}$$

$$=45272\text{N}$$

Edge distance,

$$=\frac{60}{2}=30\,\text{mm}$$

$$K_b=\frac{R}{3d_0}=\frac{30}{3\times 22}=0.4545$$

Strength of one bolt in bearing

$$=\frac{2.5\,K_b\,dt\,f_u}{1.25}$$

$$=\frac{2.5\times 0.4545\times 20\times 16\times 410}{1.25}$$

$$=104353\text{ N.}$$

Bolt Value = 45272.

No. of bolts required $=\frac{16812.5}{45272}=0.37\approx 1.$

Hence let us provide one bolt for the connection at each end of the lacing.

Check for Strength of Lacing:

$$\frac{KL}{r}=132.99$$

Referring to Table of IS800-2007, flat section belongs to buckling class 'c'.

Refer Table C of IS800-2007.

$$f_{cd}=71.8\text{ MPa.}$$

Load carrying capacity in compression of Lacing,

$$=f_{cd}\times \text{Area}$$

$= 71.8 \times 60 \times 14$

$= 60379\text{N}.$

Force in Lacing $= \dfrac{16812.5}{\sin 45^\circ} = 23776\text{ A} < 60379\text{ A}$

Hence Safe.

Let us provide 60 ISF 14 flats @ 45° and connect them to center of gravity of channel is an bolt of 20 mm nominal diameter at each end.

4. A suitable slab base for a column section ISHB 400 @ 822 N/m supporting an axial load of 500 kN. The base plate is to rest on a concrete pedestal of M20 grade concrete.

Solution:

Step 1: IS 800:2007.

Bearing Strength of Concrete Pedestal = $0.45\ f_{ck}$

$\Rightarrow 0.45\ (20)$

$\Rightarrow 9\text{ MPa}.$

Assuming a load factor 1.5; factored load,

$P_u = 1.5(500)$

$= 750\text{ kN}.$

Step 2: Area of base plate.

Area of base plate required,

$= \dfrac{750 \times 1000}{9}$

$= 83333.33\text{ mm}^2.$

The properties of ISHB 400 @ 822 N/m,

$D = 400$

$B = 250$

$t_f = 12.7$

$t_w = 10.6.$

Provide an equal projection of 40 mm on all the 4 edges of the column.

Size of base plate provided = $(400 + 2(40)) \times (250 + 2(40))$

$\Rightarrow$ 480 mm × 330 mm.

Area of plate provided,

= 158900 mm^2

= 83333.33 mm^2.

Step 3: Thickness of base plate.

Actual pressure of base,

$$= \frac{750 \times 10^3}{480 \times 330}$$

W = 4.73 MPa.

$$t_s = \sqrt{2.5\, w\left(a^2 - 0.3 b^2\right)\gamma_{mo} / f_y} > t_f$$

W = 4.73 MPa, a = b = 40 mm, $\gamma_{mo} = 1.1$, fy = 250 MPa.

t_f = 12.7 mm for given ISHB 400

$$t_s = \sqrt{2.5(4.73)\left(40^2 - 0.3(40)\right)(1.1) / 250}$$

= 7.6 > 12.7 mm

7.6 mm should be t_f.

Hence let us provide a thickness of 14 mm for the base plane.

Step 4: Connection.

Let us use 4 bolts of 20 mm diameter and 300 mm long to anchor the plate of size.

480 × 330 × 14 mm to the concrete foundation welds.

The length available for welding,

= 2[250 + (250 – 12.7) + 400 – 2(7.5)

= 1724.6 mm

= 1724 mm.

Strength of Weld/mm,

$$= \frac{410\sqrt{3}}{1.25}$$

= 189.38 MPa.

Effective area of weld = 0.7(t) (L_e).

Equating the strength of Weld to factored axial load,

$$0.7(t)\,(L_e)\,(189.38) = (750) \times 10^3$$

$$t = \frac{750 \times 10^3}{0.7 \times 189.38 \times 1724}$$

$$t = 3.28 \text{ mm.}$$

Hence let us provide 5 mm as the size of weld.

5. Let us discuss a suitable slab base for a columns section ISHB 300, subjected to a load of 450 kN. The base plate is to rest on a concrete pedestal of M 20 grade.

Solution:

Total column load = 900 kN

Bearing strength of concrete = 400/mm^2

Area of base plate $= \frac{900 \times 1000}{4} = 2{,}25{,}000 \text{ mm}^2.$

The base plate dimensions should be so chosen that the projections beyond the overall dimensions are nearly equal.

Suppose the base plate be = 530 mm × 425 mm

Area of base plate = 530 × 425 = 2, 25, 250 mm^2

Projections of the base plate beyond the all dimensions 250mm × 350 mm are,

$$\frac{425-250}{2} = 87.5 \text{ mm} = b$$

$$\frac{530 \quad 350}{} = 90\text{mm} = a$$

Upward pressure on the plate,

$$\omega = \frac{950 \times 1000}{530 \times 425} \text{ N/mm}^2$$

$$\omega = 3.995 \text{ N/mm}^2.$$

Thickness of the base plates,

$$t = \sqrt{\frac{S\omega}{\sigma_{bs}}\left(a^2 - \frac{b^2}{4}\right)}$$

$$= \sqrt{\frac{3\times3.995}{185}\left(90^2 - \frac{87.5^2}{4}\right)}$$

$= 120$ mm.

Hence the base will be 530 mm × 425 mm × 20 mm.

Let us also prove dc nominal angles to connect the column flange and the base plate.

3.5 Gusseted Base

Gusseted bases are provided for columns carrying heavier loads requiring large base plates. A gusseted base consists of a base of reduced thickness and two gusseted plates are attached one to each flange of the column.

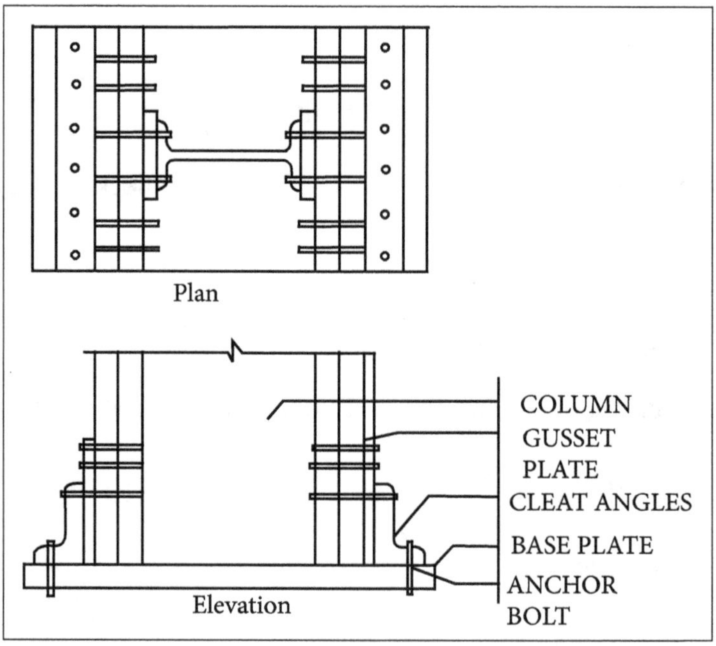

Gusseted Column Base.

The gusseted plates, cleat angles and fastenings (bolts, rivets) in combination with bearing area of shaft shall be sufficient to take all loads.

Problems

1. Let us discuss a guaranteed base for a column ISHB 3Ω @ 9.10 N/m with two plates 450 mm × 30 mm carrying a factored load of 3600 kN. The column is to be supported on concrete pedestal to be built with M20 concrete.

Solution:

$$f_{ck} = 20 \text{ N/mm}^2$$

$$A = \frac{P_u}{0.45 f_{ck}} = \frac{3600 \times 10^3}{0.45 \times 20}$$

$$A = 400000 \text{ mm}^2$$

Selecting ISA 150115, 15 mm angle and ub mm thick gusset plate.

Minimum width required,

$$= 350 + 2 \times 20 + 2 \times 16 + 2 \times 115$$

$$= 652 \text{ mm.}$$

Use 700 mm wide plate.

$$\text{Length of base plate} = \frac{400000}{700} = 571\text{mm}$$

Provide 700 × 600 mm plate.

Pressure under base plate,

$$= \frac{3600 \times 10^3}{700 \times 600} = 8.57 \text{ N/mm}^2$$

$$a = 700 - \frac{(350 + 20 \times 2 + 16 \times 2 + 2 \times 15)}{2}$$

$$a = 124 \text{ mm.}$$

BM at section X-X per mm width

$$= 8.57 \times \frac{124^2}{2} = 65886 \text{ N} - \text{mm.}$$

At section Y-Y, bending moment.

[Note: Per mm width P = 8.57 × 350 N],

$$H_{gv} = 8.57 \times \frac{350^2}{2} - \frac{700}{2} \times 8.52 \times \left(\frac{350}{2} + 20 + \frac{16+15}{2}\right)$$

$$= 106482 \text{ N-mm.}$$

Design moment = 106482 N-mm.

$$\text{Bending strength} = \frac{b}{_{mo}} = \frac{250}{1.1} = 227.27 \text{ N/mm}$$

Equating moment of resistance to bending moment we get,

$$1.2\times\frac{1}{6}\times1\times t^2\times227.27=106482$$

t = 48.4 mm.

Use 56 mm base plate of size 700 × 600 mm.

Assuming ends of columns are faced for complete bearing, the connection between gusset plate and column will be designed for 50 percent of axial load.

Design load = 0.5 × 3600 = 1800 kW

Using 24 mm shop bolts,

Strength of bolt in single shear,

$$=0.78\times\frac{\pi}{4}\times24^2\times\frac{400}{\sqrt{3}}\times\frac{1}{1.25}$$

= 65192 N

Strengthening bearing is higher.

Bolt Value = 65192 N,

No. of bolts required = $\frac{900\times10^3}{65192}=13.8$

2. Let us design a Steel column of effective height 5m has to transmit an axial load of 2000KN. Design the gusseted base. The size of the column is ISHB450. Adopt M20 concrete.

Solution:

Bearing Strength of concrete=0.6fck = 12N/mm²

Area Required= $\frac{2000\times10^3}{12}=166.7\times10^3$

Length of base plate+ Column Depth+ 2 Cover plate thickness + 2 gusset plate thickness + 2 gusset plate + 2 gusset angle + 2 Projections = 450+2(06)+2(16)+2(115)+2(25)

= 774mm say 780mm.

Breadth of plate = $\frac{166.7\times10^3}{780}$ =213.7mm say 400mm 780

Use a base plate of 780x400mm

$$\text{Upward pressure} = \frac{2000 \times 10^3}{780 \times 400} = 6.4\text{N/mm2}$$

Design of Thickness of Base Plate

$$\text{Projection} = \frac{25+780-774}{2} = 28\,\text{mm}$$

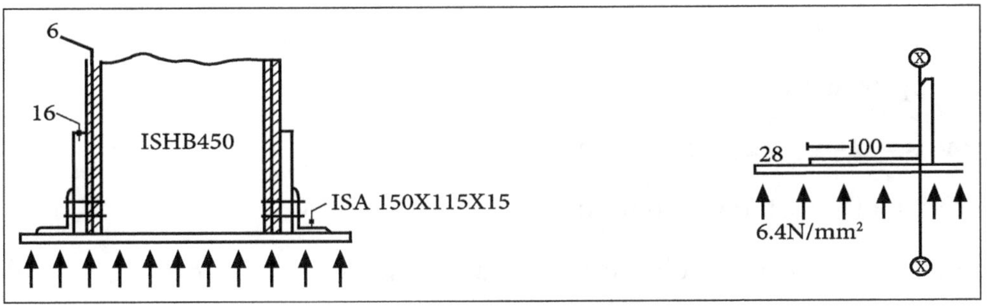

$$M_{xx} = p\frac{x^2}{2}$$

$$= \frac{6.4(100=28)^2}{2}$$

$$M_{xx} = 52428.8\ \text{N}-\text{mm}$$

$$52428.8 = 1.2 \times \frac{1}{6} \times 1 \times t^2 \times \frac{250}{1.1}$$

$$t = 33.96\text{mm}$$

$$t\ \text{plate} = 33.96 - 15$$

18.96 say 20 mm.

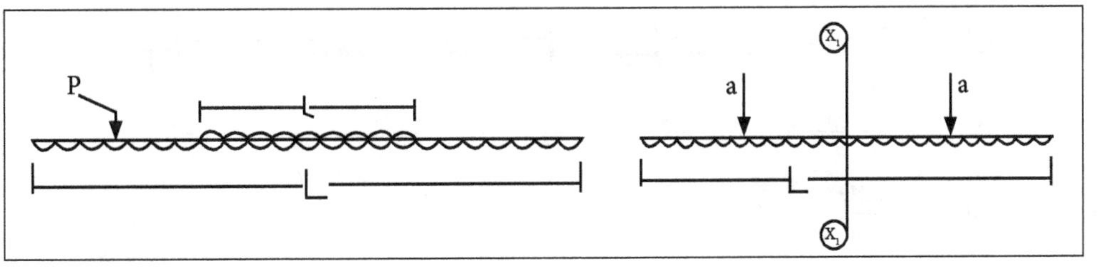

$$M_{xx} = p\frac{(I-2a)^2}{8} - p\frac{a^2}{2}$$

a = projection + angle width

a = 28+115 = 143mm.

$$M_{xx} = \frac{6.4\,(780-2\times143)^2}{8} - \frac{(6.4\times143)^2}{2}$$

$$M_{xx} = 129792$$

$$129792 = 1.2\,\frac{1}{6}\times1\times t_1^2\times\frac{250}{1.1}$$

$$t_1 = 53.44\text{ mm.}$$

Due to approximation in boundary conditions.

Thickness = 0.5 x 53.44= 26.72mm say 28mm.

Connection of gusset to the column.

Assuming the ends of the column are machine cut,

$$V_{dsb} = \frac{f_u}{\sqrt{3}\gamma_{mb}}\times\frac{\pi}{4}\times(d_b)^2$$

Using 16 mm dia bolts,

$$V_{dsb} = \frac{410}{\sqrt{31.25}}\times\frac{\pi}{4}\times\frac{16^2}{1000} = 38\,KN$$

t_f = 16mm ; thickness of gusset plate = 16mm.

$$V_{dpb} = 2.5kb \times dt \times fu$$

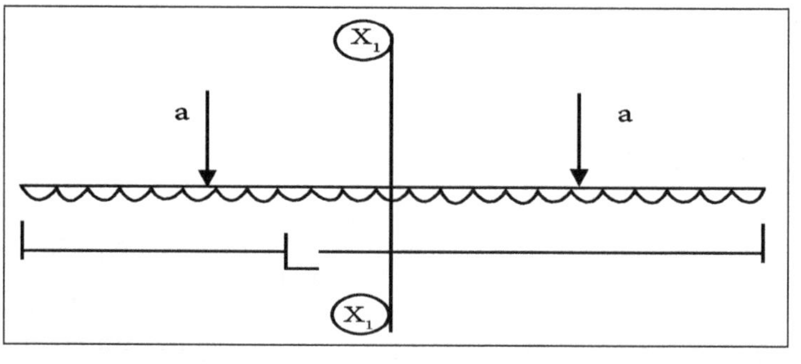

Number of bolts = $\frac{500}{38}$ =13.15.

4

Beams

4.1 Design of Laterally Supported and Unsupported Beams

Laterally Unsupported Beam

When the compression flange is not supported, it has a tendency to bend in the lateral dissection with twisting. This is called laterally unsupported beam.

Example: flange cleats.

Design Steps for Laterally Unsupported Beams

The design of laterally unsupported beams consists of selecting a section based on the plastic section modulus and checking for its shear capacity, deflection, web buckling and web crippling. Most of the equations are available in IS 800: 2007.

All the steps given in design of laterally supported beams shall be used here. Also the plastic section modulus required is increased by 25 - 50%. The design bending strength is calculated using the appropriate provisions in the code for lateral supported beams. Other checks like deflection, shear and local criteria will be same.

Web Buckling

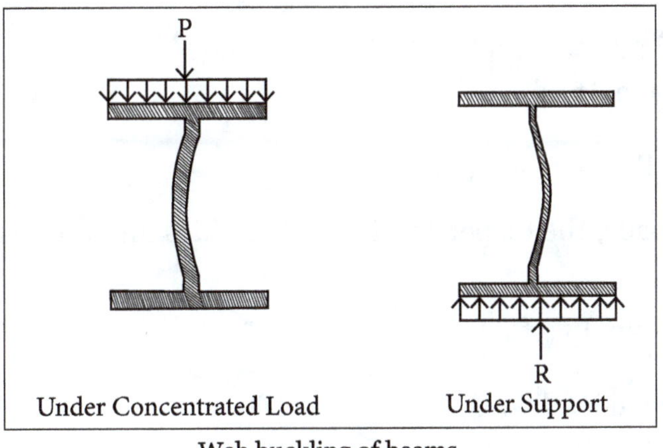

Web buckling of beams.

At point of concentrated load and at supports, un-stiffened webs of universal beam and compound beams are likely to fail by buckling.

The web of the beam is thin and can buckle under reactions and concentrated loads with the web behaving like a short column fixed at the flanges. The unsupported length between the fillet lines for I sections and also the vertical distance between the flanges or flange angles in built up sections will buckle as a result of reactions or concentrated loads. This is called web buckling.

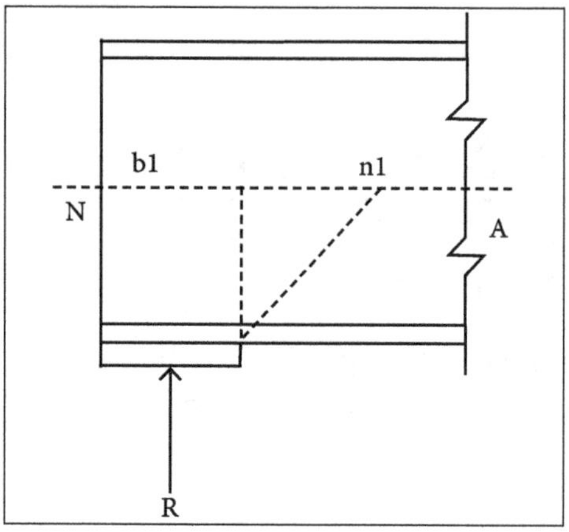

Web buckling calculation.

For safety against web buckling, the resisting force shall be greater than the reaction or the concentrated load. It will be assumed that the reaction or concentrated load is dispersed into the web at 45° as shown in the figure:

- Let Resisting force = F_{wb}.
- Thickness of web = t_w.
- Design compressive stress in web = f_{cd}.
- Width of bearing plate = b_1.
- Width of dispersion = n_1,

$$F_{wb} = (b_1 + n_1)t_w f_{cd}$$

$\geq$ Reaction, RU

For concentrated loads, the dispersion is on both sides and the resisting force can be expressed as,

$$F_{wc} = \left[(b_1 + 2n_1)\, t_w\, f_{cd}\right]$$

$\geq$ Concentrated load, WU

The design compressive stress f_{cd} is calculated based on a effective slenderness ratio of,

$0.7\, d / r_y$.

Where,

d = clear depth of web between the flanges.

r_y = radius of gyration about y-y axis and is expressed as,

$$= \sqrt{(I_{yy} / \text{area})} = \sqrt{\left[(t_w)^3 / 12\right] / t} = (t_w)^2 / 12$$

$$k_1 / r_y = (0.7d) / \sqrt{(t_w)2 / 12]} = 2.425 * d / t_w$$

Web Crippling or Crimpling

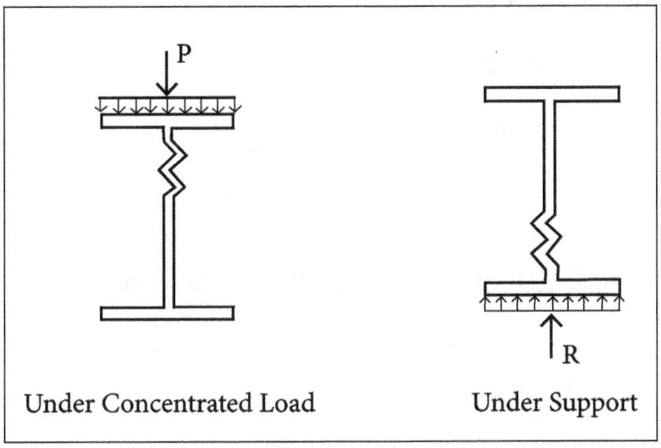

Web crippling of beams.

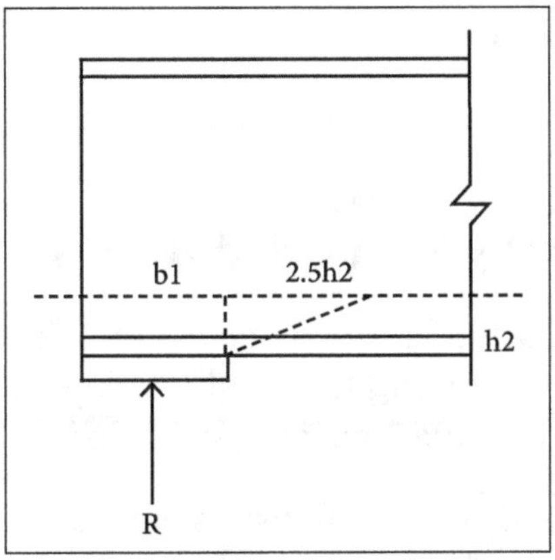

Web crippling calculations.

Web crippling causes local crushing failure of web because of large bearing stresses under reactions at supports or concentrated loads. This happens due to stress concentration because of the bottle neck condition at the junction between flanges and web. It is due to the large localized bearing stress caused by the transfer of compression from relatively wide flange to narrow and thin web. Web crippling is the crushing failure of the metal at the junction of flange and web. Web crippling causes local buckling of web at the junction of web and flange.

For safety against web crippling, the resisting force shall be greater than the reaction or the concentrated load. It will be assumed that the reaction or concentrated load is dispersed into the web with a slope of 1 in 2.5 as shown in the figure:

- Let Resisting force = F_{wc}.
- Thickness of web = t_w.
- Yield stress in web = f_{yw}.
- Width of bearing plate = b_1.
- Width of dispersion = n2 = 2.5 h_2.
- Depth of fillet = h2 (from SP [6]),

$$F_{wc} = \left[(b_1 + n_2)\, t_w\, f_{yw}\right] / \gamma_{mo}$$

≥ Reaction, R_U

For concentrated loads, the dispersion is on both sides and the resisting force can be expressed as,

$$F_{wc} = \left[(b_1 + 2\, n_2)\, t_w\, f_{yw}\right] / \gamma_{mo}$$

≥ Concentrated load, W_U

Laterally Restrained Beam

A beam in which the compression flange is restrained laterally is called a laterally restrained beam.

Design Steps for Laterally Supported Beams

The design of laterally supported beams consists of selecting a section based on the plastic section modulus and checking for its shear capacity, deflection, web buckling and web crippling. Most of the equations are available in IS 800: 2007. The steps are:

- The maximum BM and SF at collapse is calculated based on the service loads (characteristic loads) and the span of the beam. Factored load at collapse = 1.5 * characteristic loads.
- The plastic section modulus, ZPZ is calculated using:
  - $Z_{PZ} = M_U / (fY / 1.1)$.
  - M_U = Maximum BM.
  - f_Y = Yield stress of the given grade of steel.

- ◦ = Partial safety factor in yielding.
- A trial section having the appropriate plastic section modulus is adopted using IS 800 or SP (6) depending upon the type of section required. The section shall be plastic or compact section.
- The beam shall be checked for shear lag and design bending strength as given in cl. 8.2.1.5 and 8.2.1.2 (pp -53).
- The beam is checked for deflection using appropriate formula depending on the type of loadings.
- The section is checked for shear as given in cl. 8.4, 8.4.1 and 8.4.1.1. If $VU \geq 0.6$ Vd it is a case of high shear or otherwise low shear. For high shear, the design bending strength is calculated from cl. 9.2.
- The section is checked for web buckling and crippling using appropriate formula.

Deflection of a Beam

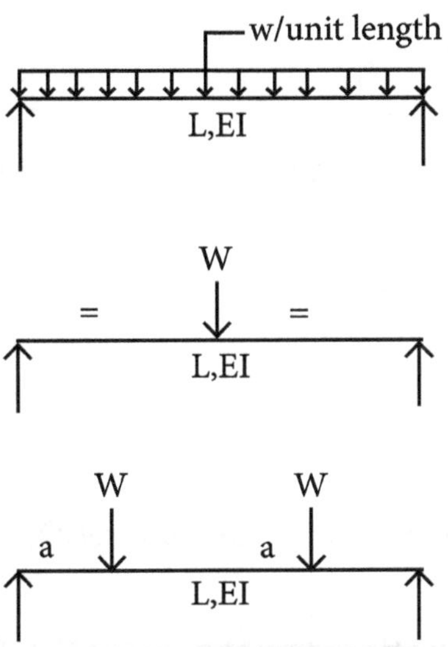

$\delta=5/384*(wL^4/EI)$

$\delta=WL^3/(48EI)$

$\delta=[WL^3/(24EI)]*\beta*(3-4\beta^2)$

$\beta=a/L$

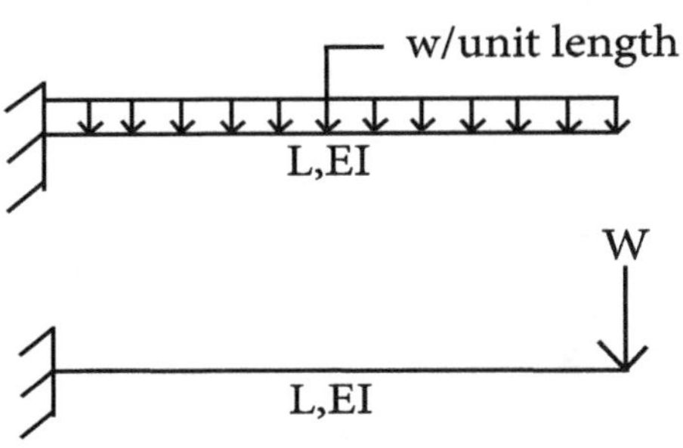

$\delta=wl^4/8EI$

$\delta=WL^3/3EI$

Problems

1. Let us discuss the problem based on a simply supported beam of 10 m effective span carrying a total factored load of 60 kN/m. The depth of beam should not exceed 500 mm. The compression flange of the beam is laterally supported by floor construction. Assume stiff end bearing is 75 mm.

Solution:

$$L = 10 \text{ m} = 10000 \text{ mm}, w = 60 \text{ kN/m}.$$

Trial Section:

Maximum BM,

$$M = \frac{wL^2}{8} = \frac{60 \times 10^2}{8} = 750 \text{kN/m}$$

$\therefore Z_p$ required

$$= \frac{M\gamma_{mo}}{f_y} = \frac{750 \times 10^6 \times 1.1}{250} = 3300 \times 10^3 \text{ mm}^3.$$

Since depth restricted is 500 mm, select ISMB 450 and suitable plates over flanges,

Z_p of ISMB 450 = 1553.4×10^3

Z_p to be provided by cover plates

$= (3300 - 1553.4) \times 10^3$

$= 1746.6 \times 10^3 \text{ mm}^3$.

If A_p is the area of cover plate on each side tensile force an compressive forces developed at the time of hinge formation = $A_p f_y$.

If the distance between the two plates is 'd', plastic moment resisted = $A_p f_y d$.

Hence the additional Z_p provided by the cover plates may be taken as,

$$Zp \text{ of plates} = \frac{A_p f_y d}{f_y} \times \frac{1}{\gamma_{mo}} = \frac{A_{pd}}{1.1}$$

$$\therefore \frac{A_{pd}}{1.1} = 1746.6 \times 10^3.$$

Taking d = 450 + t = 450 mm.

We get,

$$A_p = \frac{1746.6 \times 10^3 \times 1.1}{450} = 4269.5\ \text{mm}^2$$

Provide 220 × 20 mm plates on either side.

Check for shear,

$$V_d = \frac{f_y}{\sqrt{3}} \times \frac{1}{\gamma_{mo}} \times h \times t_\omega = \frac{250}{\sqrt{3}} \times \frac{1}{1.1} \times 450 \times 9.4$$

$$= 555.044 \times 10^3\,\text{N}$$

$$V = 60 \times \frac{10}{2} = 300\ \text{kN}.$$

Check for Moment Capacity:

Section classification: Outstanding element of compression flange,

$$\frac{b}{t_b} = \frac{150/2}{17.4} = 4.3 < 9.4\ \text{E}$$

$$\frac{d}{t_w} = \frac{450 - 2(19.4 + 15)}{9.4} = 40.9 < 84\text{E}$$

Hence plastic section,

$$H_d = 1.0 \times \frac{Z_p f_y}{\gamma_{mo}} < 1.2\ Z_e\, f_{hy} \frac{1}{\gamma_{mo}}$$

$$\frac{Z_p f_y}{\gamma_{mo}} = \left[1553.4 \times 10^3 + 220 \times 20(450 + 20)\right] \times \frac{250}{1.1}$$

$$= 823.0455 \times 10^6\ \text{N} - \text{mm}$$

$$I_{zz} = 789.888 \times 10^6 \text{mm}^4$$

$$Z_e = \frac{789.88 \times 10^6}{255 + 20} = 3224.033\ \text{mmm}^4$$

$$M_d = 823.043 \times 10^6 > H.$$

Hence O.K.

Check for Deflection

$$\text{Working load} = \frac{60}{1.5}\,\text{kN/m} = 40\ \text{kN/m} = 40\ \text{N/mm}$$

$$\delta = \frac{\delta \omega L^4}{384\, EI_{ZZ}} = \frac{5 \times 40 \times (10000)^4}{384 \times 2 \times 10^5 \times 789.884 \times 10^6} = 32.97\ \text{mm}$$

Permissible, if elastic cladding is assured,

$$(\text{Table of IS 800}),\ \frac{L}{240} = \frac{10{,}000}{240} = 41.67\ \text{mm}$$

Hence safe.

Check for Web Buckling

$$h = 450\ \text{mm}.$$

$$\therefore \lambda = 2.5\frac{h}{t_w} = 2.5 \times \frac{450}{9.4} = 119.68$$

From Table of IS 800-2007,

$$f_{cd} = 94.6 - \frac{9.68}{10}(94.6 - 83.7) = 84.05\ \text{N/mm}^2$$

$$\therefore F_{cdw} = (b_1 + n_1)t_w f_{cd} = \left(75 + \frac{450}{2}\right)9.4 \times 84.05$$

$$= 237.021 \times 10^3\,\text{N} < 300\ \text{N}.$$

Hence needs increased effective bearing length b. Provide b = 175 mm.

Then,

$$F_{cdw} = \left(175 + \frac{450}{2}\right) \times 0.4 \times 84.05$$

$$= 316.028 \times 10^3 > 300\ \text{kN}.$$

∴ Safe.

Check for Web Crippling

$$F_w = (b_1 + 2.5t_f)f_y \frac{1}{\gamma_{mo}} t_w$$

$$= [175 + 2.5 \times (17.4 + 20)]\,250 \times \frac{1}{1.1} \times 9.4$$

$= 573.614 \times 10^3 N$? 300 kN.

Hence safe.

Design of Connection between Flange Plates and Flange:

Bolts/welds joining the plates and flange are to be designed for the horizontal shear at that level. Shear stress at the level of plates and flanges,

$$= \frac{F}{bI_{ZZ}}(\bar{a}_y) = \frac{300 \times 10^3}{150 \times 789.884 \times 10^6}(220 \times 20 \times 225)$$

$$= 2.506 \text{ N/mm}^2.$$

2. A simply supported laterally restrained beam of effective span 4m carrying a factored point load of intensity 50kN at the mid span. Let us design an appropriate section using Fe410 grade steel.

Solution:

Step 1: Calculation of BM and Choosing a Trial Section.

Maximum B.M,

$$M = \frac{\omega l}{4} = \frac{50 \times 4}{4} = 50 \text{ KN.m.}$$

$$M = 50 \times 10^6 \text{ N.mm.}$$

Maximum Shear,

$$= \frac{\omega l}{2}$$

$$= \frac{50 \times 4}{2}$$

$$= 100 \text{ kN.m.}$$

$$Z_P \text{ required} = \frac{M\gamma_{mo}}{f_y} = \frac{50 \times 10^6 \times 1.1}{250}$$

$$= 2.2 \times 10^6 \text{ mm}^3.$$

ISMB 550 with suitable plates over the flanges.

Step 2: Choosing Cover Plates.

Z_p of ISMB 400 = 11.76, 20 × $10^3 mm^3$

Z_p to be provided by the cover plate,

= 2.20 × 10^6 – 1.176 × 10^6

= 1.0 × $10^6 mm^3$.

Let A_p be the area of cover plate on each side of the flange,

$E = 2 \times 10^5 \; N/mm^2$

$Z_R = 2359.8 \; cm^3 = 2359 \times 10^3 \; mm^3$

$Z_p = 2711.08 \; cm^3 = 2711.98 \times 10^3 \; mm^3$

$D = 550 \; mm$

$b_f = 190 \; mm$

$t_f = 19.3 \; mm$

$t_\omega = 11.2 \; mm$

$f_y = 250 \; N/mm^2 \; r_1 = 18mm$

$$\varepsilon = \sqrt{\frac{250}{f_y}} = 1$$

$$b = \frac{b_f}{2} = \frac{190}{2} = 95 \, mm$$

$$\frac{b}{t_f} = \frac{95}{19.3} = 4.92 < 9.4\varepsilon \; \therefore \text{Plastic Section.}$$

$$d = D - 2t_f - 2r_1$$

$$= 550 - 2[19 - 3] - 2[18]$$

$$d = 475.4 \; mm$$

$$\frac{d}{t_\omega} = \frac{475.4}{11.2} == 42.45 \rightarrow 84\varepsilon \; \therefore \text{Plastic Section.}$$

∴ Section is classified as plastic section.

Design Shear Strength of the Section:

[as per Cl.8.4 of IS 800:2007],

V = 100kN.

$$V_d = \frac{V_n}{\gamma_{mo}}$$

$$V_n = \frac{A_v f_{y\omega}}{\sqrt{3}}$$

$$= \frac{[ht_\omega]f_y}{\sqrt{3}} - \frac{550 \times 11.2 \times 250}{\sqrt{3}}$$

$$V_n = 889 \text{ kN.}$$

$$V_d = \frac{889}{1.1} = 808 > 100 \text{ kN.}$$

Hence Safe.

Design Bending Strength of the Section:

The section is plastic and laterally supported,

$$M_d = \frac{\beta_b Z_p f_y}{\gamma_{m_o}} \leq \frac{1.2 Z_e f_y}{\gamma_{m_o}}$$

$$\beta_b = 1 \text{ for Plastic Section}$$

$$Z_p = 2711.98 \times 10^3 \text{mm}^3$$

$$Z_e = 2359 \times 10^3 \text{ mm}^3$$

$$= \frac{1 \times 2711.98 \times 10^3 \times 250}{1.1} \leq \frac{1.2 \times 2359 \times 10^3 \times 250}{1.1}$$

$$= \frac{616 \text{ kN.m} \leq 643.3 \text{ kNm}}{\text{series sake}}$$

$$= 1.0 \times 160^\circ \text{mm}^2.$$

Let Ap be the area of cover plate on each side of the flange.

Limiting Deflection,

$$y = \frac{\omega L^3}{48EI} \leq \frac{l}{300}$$

$$= \frac{50 \times 10^3 \times 4000^3}{48 \times 2 \times 10^5 \times 64893.34 \times 10^4} \leq \frac{4000}{300}$$

$$y = 0.513 \leq 13.33 \text{ mm.}$$

Hence Safe.

3. Let us design a laterally unsupported beam of 4m effective span, carrying a factored bending moment of 350 kNm and factored shear force of 100 kN. Use Fe410 grade steel.

Solution:

- Design Laterally Unsupported Beam.
- Effective Span = 4m.
- Factored B.M = 350kN.m.
- Factored S.F = 100kN.
- Use Fe410 grade steel.

  ⇒ Choose ISMB i.e., 550

  $Z_p = 2711.9 \times 10^3 \ 10 / mm^3 / Z_e \ 2359 \times 10^3 mm^3$

  $D = h = 550$ mm

  $b_f = 190$ mm

  $t_f = 19.3$ mm

  $t_w = 11.2$ mm

  $\gamma_1 = 221.6$ mm

  $f_y = 410 \ N / mm^2$

  $$\varepsilon = \frac{b_f}{2} = \frac{190}{2} = 95 \text{mm}$$

$$\frac{b}{t_f} = \frac{95}{19.3} = 4.92 \leq 9.4\varepsilon [\text{Plastic Section}]$$

$$d = D - 2[t_f + \gamma_1]$$

$$= 550 - 2\ [19.3 + 18]$$

$$d = 475.4\ \text{mm}$$

$$\frac{d}{t_w} = \frac{1475.4}{11.2} = 42.96 \leq 84\varepsilon\ [\text{Plastic Section}].$$

f_{crb} :

$$KL = 4000\text{mm}.$$

$$\frac{K^2}{\gamma_{xx}} = \frac{4000}{221.6} = 18.05$$

$$\frac{b}{t_f} = \frac{550}{19.3} = 28.20.$$

Refer Table of IS 800, for $\frac{KL}{r} = 18.05$ & $\frac{h}{t_f} = 28.2$,

$$f_{crb} = 7.4\ \text{N/mm}^2$$

f_{bd} :

$\alpha LT = 0.21$ for rolled section.

Refer Table for $\alpha LT = 0.21$, $f_y = 40\ \text{N/mm}^2$,

$$f_{bd} = 374.1\ \text{N/mm}^2$$

Design bending strength,

$$M_d = \beta_b\ z_P\ f_{bd}.$$

$$\beta_b = 1\ [\text{for Plastic Section}]$$

$$Z_p = 2711.9 \times 10^3\ \text{mm}^3$$

$$M_d = 1 \times 2711.9 \times 10^3 \times 216.9$$

$$M_d = 588.23 \times 10^6 \text{ N/mm} \ \& \ 350 \times 15^6 \text{ N.mm.}$$

Design Shear Strength of the Section:

$$V_d = \frac{V_A}{\gamma_{my}}, V_n = \frac{A_V f_{yw}}{\sqrt{3}}$$

$$V_n = \frac{ht_w f_{yw}}{\sqrt{3}} = \frac{550 \times 11.2 \times 410}{\sqrt{3}}$$

$$V_n = 1458.15 \text{ kN}$$

$$V_d = \frac{1456.15}{1.1} = 1325.5 \text{ kN} > 100 \text{ kN.}$$

Hence Safe.

4. A laterally supported beam having an affective span of 8 m consists of ISMB 550 @ 103.7 kg/m and covet plate of 250 mm × 16 mm connected to each flange by 20 mm diameter rivets. Let us determine the safe UDL which the beam can carry in addition of its own weight.

Solution:

- Effective Span = 8 m.
- ISMB 550 @ 103.7 kg/m.
- Cover Plate 250 mm × 16 mm.
- Connect to each flange 20 mm dia rivets.

⇒ Section Properties:

$$D = 550 \text{ mm.}$$

$$b_f = 190 \text{ mm.}$$

$$t_f = 19.3 \text{ mm.}$$

$$t_w = 11.2 \text{mm.}$$

$$r_1 = 22.16 \text{mm.}$$

$f_y = 250$ MPa.

$Z_p = 2711.98 \times 10^3 \, mm^3$.

$Z_e = 2359.80 \times 10^3 \, mm^3$.

Classification of Section:

$$b = \frac{b_f}{2} = \frac{190}{2} = 95 \, mm.$$

$$\frac{b}{t_f} = \frac{95}{19.3} = 4.9 < 9.4\varepsilon \text{ [Plastic Section]}$$

$$d = D - 2[t_f + r_1]$$

$$= 550 - 2[19.3 + 22.16]$$

$$d = 467.08 \text{ mm}$$

$$\frac{d}{t_w} = \frac{467.08}{11.2} = 41.7 \text{ mm} < 84\,\varepsilon \text{ [Plastic Section].}$$

To find f_{crb}: KL = L = 8000 mm,

$$\frac{KL}{\gamma_{xx}} = \frac{8000}{221.6} = 36.1$$

$$\frac{h}{t_f} = \frac{550}{19.3} = 28.49$$

Referring to Table of IS800-2007 for,

$$\frac{kL}{r} = 36.1 \,\&\, \frac{h}{t_f} = 28.49$$

$$f_{crb} = 1680.1 \text{ N/mm}^2.$$

To find f_{bd}: $\alpha L_T = 0.21$ for rolled section.

By referring 13. (a) for,

$\alpha L_T = 0.21$ & $f_y = 250$ MPa.

$f_{crb} = 1680.1$ N/mm².

$f_{bd} = 224$MPa.

Design Bending Strength:

$$M_d = \frac{\beta_b Z_P f_y}{\gamma_{mo}} \leq \frac{1.2 Z_e f_y}{\gamma_{mo}}$$

$\beta_o = 1$ for plastic section,

$$M_d = \frac{1 \times 250 \times 2712 \times 10^3}{1.1} \leq \frac{1.2 \times 2359.8 \times 10^3 \times 250}{1.1}$$

$$= 616363636.4 \leq 643581818.2 \text{ N.}$$

Factored udl for the Section:

$$M = \frac{\omega l^2}{8}$$

$$616363636.4 = \frac{\omega \times 8000^2}{8}$$

$\omega = 077.05$ N / mm Self Weight.

Design Shear Strength:

$$V_d = \frac{V_n}{\gamma_{mo}}$$

$$V_n = V_P = \frac{A_v \delta_{yw}}{\sqrt{3}}$$

$$V_n = \frac{[550 \times 11.2]250}{\sqrt{3}} = 8.89119 \text{ N.}$$

$$V_d = \frac{889119}{1.1} = 808290 \text{ N.}$$

Factored Udl on Design Shear:

$$\frac{\omega l}{2} = V_d$$

$$808290 = \frac{\omega \times 8000}{2}$$

$\omega = 202.07$N/mm + Self Weight

∴ Permissible udl 77.05 N/mm = 77.05 kN/m + Self-Weight.

5. Let us design a simply supported beam of effective span 12 m to carry a factored load 70 kN/m the depth of beam is restricted to 500 mm.

Solution:

Effective Span = 1200

Factored udl on beam = 70 kN/m.

Step 1: Calculation of B.M and Choosing a Trial Section:

Maximum,

$$BM = \frac{Wl^2}{8} = \frac{70(12)^2}{8} = 1260 \text{ kN.m}$$
$$= 1260 \times 10^6 \text{ N.mm}$$

$$ZP \text{ required} = \frac{M_{\gamma mo}}{fy} = \frac{1260 \times 10^6 (1.1)}{250}$$
$$= 5.511 \times 10^6 \text{ mm}^3.$$

Step 2: Choosing cover plate $= 1176.2 \times 10^3 \text{ mm}^3$

$$\therefore \text{Zp cover plate} = \left(5.544 \times 10^6 - 1176.2 \times 10^3\right)$$
$$= 4.368 \times 10^6 \text{ mm}^3$$
$$= (\text{Area of Plate}) \frac{fy}{\gamma_{mo}} (\text{Distance between the plate})$$
$$= \frac{A_P fy}{\gamma_{mo}} (400)$$

$$M = f_y \cdot Z_p$$

$$Z_P = \frac{A_P fy(4000)}{\gamma_{mo} fy} = \frac{A_P(400)}{1.1}$$
$$= 4.368 \times 10^6 \text{ mm}^3.$$

$$M_d = B_b \cdot \frac{Z_{pty}}{\gamma_{mo}}$$
$$= (1.0)\left[\frac{Z_P \cdot fy}{\gamma_{mo}}\right](Z_P \text{ of section} + Z_P \text{ plate})$$

$$M_d = 1176.2 \times 10^3 + (320 \times 40)\ (400 + 40) \cdot \frac{250}{1.1}$$

$$M_d = 1547.318\ \times 10^6\ N \cdot mm$$

$$Z_{fd} = [Z_P\ I\ section\ +\ Z_P\ of\ plate\ -\ A_w \cdot yw]$$

$$= 1176.2\ \times\ 10^3 + (320 \times 40)(400 \times 410) - \left(400 \times 8.9 \times \frac{400}{4}\right)$$

$$= S\left[1176.2 \times 10^3 + 5632\ \times\ 10^3 - 356 \times 10^3\right]$$

$$= 6452.2\ \times\ 10^3 mm^3$$

$$M_{dv} = M_d - \beta\ (M_d - M_{fd})$$

$$= 1547.318 \times 10^6 - 0.6371(1547.31 \times 10^6\ -\ 1466.44 \times 10^6)$$

$$= 1547.318\ \times\ 10^6 -\ 51.53\ \times\ 10^6$$

$$\Rightarrow\ 1496\ \times\ 10^6\ Nmm$$

$$\text{Working load} = \frac{70}{1.5} = 46.67\ kN/m$$

$$4.368 \times 10^6 = \frac{A_P(400)}{1.1}$$

$$A_P = 12012\ mm^2.$$

Step 3: Check for Shear:

$$V_d = \frac{(fy/\sqrt{3})}{\gamma_{mo}} ht_w = \frac{(250/\sqrt{3})}{1.10} \times\ 100 \times\ 8.9$$

$$= 467143\ N.$$

$$0.6\ V_d = 0.6(467143) = 280286\ N$$

$$S.F = \frac{W_1}{2} = \frac{70000}{2}\ (12) = 420 \times\ 10^3\ N.$$

Step 4: Section Classification:

$$\frac{b}{t_f}=\frac{140}{6}=8.75 < 29.3$$

$$\frac{d}{t_w}=400-\frac{2(t_f+r_1)}{t_w}$$

$$=\frac{400-2(16.0+14)}{8.9}$$

$$=38.20\ 24$$

Step 5: Check for B.M:

$$Mdv = Md - \beta\ (M_d - M_{fd})$$

$$\beta=\left(\frac{2V}{V_d-1}\right)^2$$

$$=\left[\frac{2(420\times10^3)}{467143}-1\right]^2$$

$$=\ 0.6371$$

$$S=\frac{5\,WS^4}{384\ EI_{ZZ}}=\frac{5\times46.67\times(12000)^4}{384\ (2\times10^5)I_{ZZ}}$$

$$I_{XX}=I_{ZZ}\text{ for ISMB }400+I_{ZZ}\text{ for plate}$$

$$=20458.4\times10^4+2\times A_P\ \left(\frac{d}{2}\right)^2$$

$$=20458.4\times4\times10^4+2\left[(320\times40)\left(\frac{400}{2}+\frac{40}{2}\right)^2\right]$$

$$=144.36\times10^7\,mm^4$$

$$S=\frac{5\times46.67\times(120000)^2}{384(2\times10^5)\times(144.36\times10^7)}$$

$= 43.64 \text{ mm}$

$= \frac{L}{240}$

$= 50 \text{ mm} > 43.64 \text{ mm}.$

Step 6: Check for Web Buckling:

$$d = 2.5 \frac{h}{tw} = 2.5\left(\frac{400}{8.9}\right)$$

$$= 112.36 \text{ mm}$$

$$f_{cd} = 94.6 \cdot \frac{(94.6 - 83.7)}{10}(2.36)$$

$$= 92.03 \text{ MPa}$$

$$E_{cdw} = (b_1 + n_1)\ tw\ (fcd)$$

$$b_1 = 70 \text{ mm}$$

$$n_2 = \frac{400}{2} = 200 \text{ mm}$$

$$tw = 8.9 \text{ mm}$$

$$f_{cd} = 92.03 \text{ MPa}.$$

$$F_{cdw} = (70 + 200)\ (8.9)(92.03) = 221148 \text{ N}.$$

$$= h_2O \times 10^3 \text{ n}$$

$$F_{cdw} = (350 + 200)(8.9)\ (92.093)$$

$$= 450.795 \times 10^3 \text{ N}$$

$$= 420 \times 10^3 \text{ N}.$$

Check for Web Crippling:

$$Fw = \frac{(b_1 + n_2)(tw)(fy)}{\gamma_{mo}}$$

$b_1 = 350$ mm

$n_2 = 2.5$ (root radius + flange thickness)

$= 2.5\ (14+16) = 75$ mm

$tw = 8.9$ m • $fy = 250$ MPa • $\gamma_{mo} = 1.1$

$$F_W = \frac{(350.175)(8.9)(250)}{1.1} = 259.65 \times 10^3 \text{ N}$$

$> 420 \times 10^3$ N.

6. A beam is simply supported over a span of 6 m. It supports one Iron beam at mid span exerting 90 kN. Let us design the beam with ISWB section with flange plates.

Solution:

Approximate deal load of the beam,

$$= \frac{L.L \times Span}{300} = \frac{30 \times 6}{300}$$

$= 0.6$ kN / m

Total load = 30 + 0.6 = 30.60 kN/m.

$$\text{Minimum B.M} = M = \frac{\omega l^2}{8} = \frac{30.6 \times 6^2}{8} = 137.7 \text{ kN/m}$$

$$\text{Minimum S.F} = S = \frac{30.6 \times 6}{2} = 91.8 \text{ kW}$$

Permissible bending stress = 0.66 × 250 = 165 N/mm²

Permissible average shear stress = 0.4 × 250 = 100 N/mm²

Section modulus required,

$$2 = \frac{M}{\sigma_{bc}} = \frac{137.7 \times 10^6}{16^5} \text{mm}^2$$

$Z = 834.5 \times 10^3$ mm³

Provide ISWB @ 56.9 Ks/m

Properties of the section overall depth 350 mm, tω = 8 mm.

$I_{xx} = 15{,}321.7 \times 10^4\ mm^4$

$Z = 887 \times 10^3\ m^3$

$h_2 = 27.25\ mm$

$h_1 = 255.5\ mm.$

Check for Shear:

Induced average shear stress $= \dfrac{S}{dt_\omega} = \dfrac{91.8 \times 10^3}{350 \times 8} N/mm^2.$

This is less than permissible shear stress.

Check for Deflection:

Maximum deflection at the centre,

$$= \frac{5}{384}\frac{\omega l^4}{EI}$$

$$= \frac{5}{384} \times \frac{30.6 \times 6^4 \times 10^9}{200 \times 15321.7 \times 7 \times 10^4} = 16.85\ mn$$

Permissible deflection,

$$= \frac{Span}{325} = \frac{6000}{325} = 18.46\,mm$$

Since the maximum deflection is less than permissible deflection, the design is safe.

Check for crippling web at support,

Assume bearing of 300 mm.

Induced bearing stress $= \dfrac{R}{t\left(a + h_2\sqrt{3}\right)} = \dfrac{91.8 \times 10^3}{8/300 + 27.25\sqrt{3}}$

$$= 33\ N/mm^2\ \ 187.5 N/mm^2.$$

Check for Buckling of the Web:

Slenderness ratio from used buckling,

$$= \frac{}{t}\sqrt{3} = \frac{295.5}{8}\sqrt{3} = 64.$$

Safe compressive stress corresponding to above ratio,

$\sigma_{ac} = 117\ N/mm^2$.

$$B = \infty + \frac{h}{2} = 300 + \frac{350}{2} = 475\ mm$$

Load bearing capacity of the web,

$$= \sigma_{ac}\ t\ B = 117 \times 8 \times 475$$

$$= 444.6\ kN.$$

But the end reaction is only 91.8 kN.

4.2 Built-Up Beams

Beams, Built-Up Teams and Plate Girder

Any flexural member designed to take up the stresses due to bending are called as beams:

- Built up beam are normally classified as beam of standard available section, to which additional member such as plate are added to increase the structural capabilities of the section or member.
- Plate girder are normally used to carry heavy load for which normal ready-made section coupled with built-up plate and flange plate connected together by flange angle by means of welding or riveting.

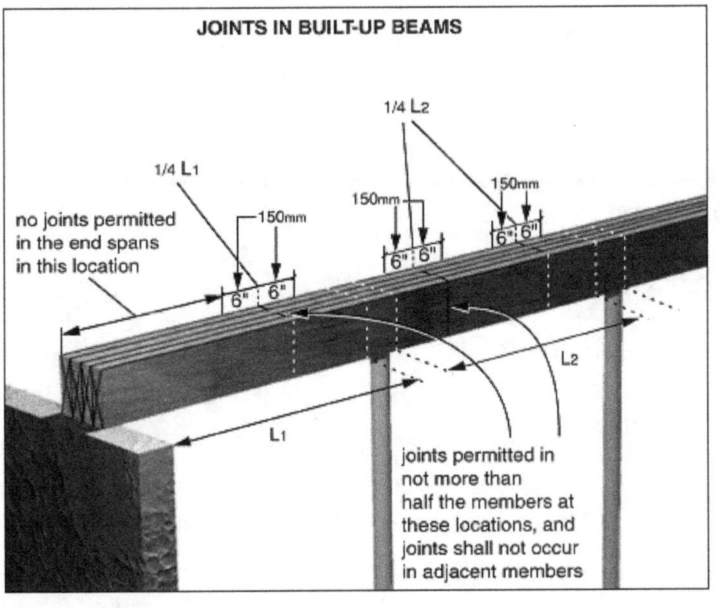

Problems

1. Let us discuss about a problem based on a built up beam station for a span 8 m to carry a uniformly distributed load 15 kN/m and central concentrated load 100 kN. The beam is laterally supported throughout, show the curtailment of plate also.

Solution:

- Permissible bearing stress = 300 N/mm2.
- Permissible shear stress = 100 N/mm2.
- Bending stress of flange plate = 165 N/mm2.

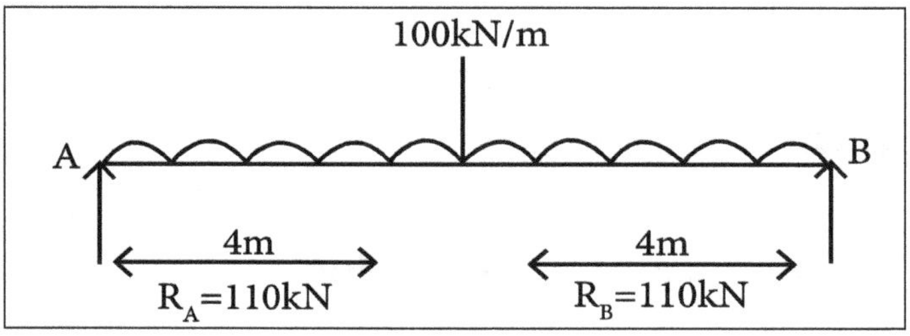

Maximum bending moment,

$$= \frac{15(8)^2}{2} + \frac{100(8)}{4}$$

$$= 320 \text{ kN} \cdot \text{m}$$

$$Z = \frac{M}{f}$$

$$= \frac{320 \times 10^6}{165}$$

$$= 1939393 \text{ mm}^3$$

$$\text{ISMB } 450 = 1350.7 \times 10^3 \text{ mm}^3$$

$$= 1939393 - 1350700$$

$$= 588693 \text{ mm}^3$$

$$A_P = \frac{Z - Z_j}{d} + \text{[Area of hole in one flange of I section]}.$$

Diagram of hole = 18 + 1.5 = 19.5 mm.

$$A_p = \left[\frac{1939393 - 1350700}{456}\right] + 2(19.5)\ (17.4)$$

$$= 1986.81 \approx 1987\ \text{mm}^2$$

$$= \frac{1987}{12} + 2(9.5) = 204.58 \text{ say } 210 \text{ mm.}$$

Check for Actual Stress

Gross moment $I = 30390.8 \times 10^4 + 2(210)\ (12)\left(\frac{450}{2} + 6\right)^2$

$$= 57285 \times 10^4\ \text{mm}^4$$

$$f_{bc} = \frac{M}{I}(y)$$

$$= \frac{320 \times 10^6}{57285 \times 10^4}\left(\frac{450}{2} + 12\right)$$

$$= 132.39\ \text{N/mm}^2$$

$$= \frac{\text{Gross area of tension flange}}{\text{Net area of tention flange}}$$

$$= \frac{(210 \times 12 + 150 \times 16) \times 132.39}{(210 \times 12 + 150 \times 17.4) - [2 \times 19.5 \times (12 + 17.4)]}$$

$$= 163.52\ \text{N/mm}^2.$$

Check for Shear

Maximum Shear Force,

$$= \frac{(15 \times 8) + 100}{2}$$

$$= 110\ \text{kN}$$

$$= \frac{S}{2t\omega}$$

$$=\frac{110\times 10^3}{450\,(9.4)}$$

$$=26\text{ N }/\text{mm}^2.$$

Curtailment of Flange Plate,

$$=f_z$$

$$=(165)\,\left(1350.7\times 10^3\right)$$

$$=222.865\times 10^6\text{N}\cdot\text{mm}$$

$$=222.865\text{ kN}\cdot\text{m}.$$

$$\left(R_A\cdot x-\frac{\omega_x^2}{2}\right)=\text{M.R}$$

$$\Rightarrow\ 110(x)-15\left(\frac{x^2}{2}\right)=222.865$$

$$\Rightarrow -7.5\,X^2+110x\ -\ 222.865=0$$

$$\Rightarrow\ 7.5\,x^2-110x+\ 222.865=0$$

$$=\frac{222.865}{320}(110)$$

$$=76.61\text{ kN/m}^2$$

$$=76.61\text{ N/mm}^2$$

$$=76.61\,(210)\,(12)=193\times 10^3\text{ N}$$

$$=\frac{\pi d^2}{4}f_s$$

$$=\frac{(3.14)(19.5)^2}{4}(100)$$

$$=29850\text{ N}$$

$$=dt\,f_b$$

Strength of one = 19.5 (12) (300)

Bearing = 70200 N

$$\phi = 29850 \text{ N}$$

No. of rivets $= \dfrac{193 \times 10^3}{29850} \approx 7 \text{ rivets}$

$$\frac{7}{2}(60) = 210 \text{ mm} = 0.21 \text{ m.}$$

∴ Actual distance from the support at which the flange plate can be curtailed,

$= 2.51 - 0.21$

$= 2.30$ m

$= 110 - 15(2.51)$

$= 72.35$ kN.

Horizontal shear per m length,

$$= \frac{Fa\bar{y}}{I}$$

$$= (210 \times 12)\left(\frac{450}{12} + \frac{12}{2}\right)$$

$$a\bar{y} = 582126 \text{ mm}^3$$

$$= \frac{(72.35 \times 10^3)(582120)}{57285 \times 10^4}$$

$$= 73.52 \text{ N/mm}^2.$$

Least rivot 18 mm dia =29850 N,

$$= \frac{2(29850)}{73.52} = 812 \text{ mm}$$

Permissible maximum pitch,

$= 125$

$= 12(12)$

$= 144$ mm self-140 mm.

∴ Hence provide a pitch of 140 mm c/c.

4.3 Beams Subjected to Uniaxial and Biaxial Bending

Beam Column

A structural member subjected to varying amounts of both axial compression and bending moment is commonly termed as a beam column.

Castellated Beam

A beam with a number of regular openings in its Web is called a castellated beam. A castellated beam is formed by flame cutting a single rolled wide flanged beam in a definite pattern and then rejoining the segments in such a way as to form a beam with number of regular opening in its web.

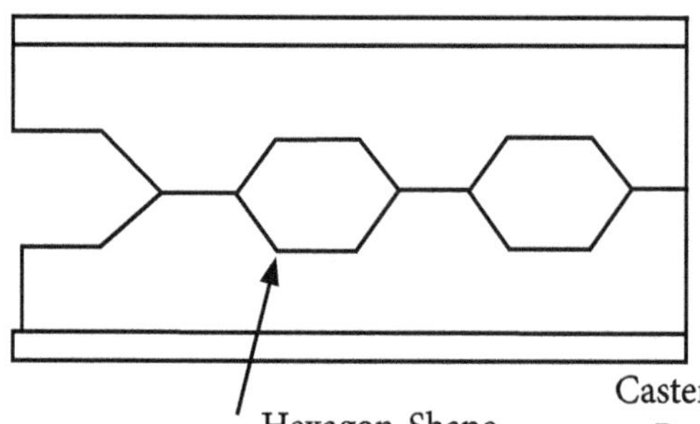

Castellated Beam.

Biaxial Bending

Many columns are subjected to biaxial bending that is bending about both axes. Corner columns in buildings where beams and girders frame into the columns from both directions are the most common cases but there are others such as where columns are cast monolithically as part of frames in both directions or where columns are supporting heavy spandrel beams. Bridge piers are almost always subject to biaxial bending.

Circular columns have polar symmetry and thus the same ultimate capacity in all directions. The design process is the same therefore regardless of the directions of the moments. If there is bending about both the x and y axes the biaxial moment can be computed by combining the two moments or their eccentricities as follows:

$$M_u = \sqrt{(M_{ux})^2 + (M_{uy})^2}$$

or,

$$e = \sqrt{(e_x)^2 + (e_y)^2}$$

4.4 Design of Plate Girders

Plate Girder and Its Uses

Building up I-section with two flanges plates connected to web plate of required depth. This type of I-beams are known as "Plate Girder".

In road a rail way bridges, to carry crane beam plate girder are used.

The Necessity of Curtailment of Flange Plate in the Plate Girder

The value of bending moment is maximum at the centre and it gradually reduces to zero in case of simply supported beams. Hence larger cross section is required at the mid span.

The cross section could be reduced to just meet the required bending moment which could help in reducing the cost of the plate girder.

Factors Governing Flange Curtailment in Plate Girders

Often the plate girder is subjected to loading then the maximum bending moment at one section usually when the plate girder is simply supported at the ends and subjected to the uniformly distributed load (Udl) then the maximum BM occurs at the flange of plate girder is designed to resist the minimum BM. The flange are designed to resist the maximum BM is not required at other section. The flange area reduced as the BM decreases. The reduction in flange area is done by curtailing the flange plate.

The function of stiffness in plate girder.

Stiffeners are normally classified as vertical and horizontal stiffeners. Vertical stiffeners are used to transmit the concentrated load to the web plate. Horizontal stiffeners are provided in case of web plate of larger heights to avoid web buckling.

Problems

1. Let us discuss about a problem based on a welded plate girder of span 24 m to carry super-imposed load of 35 kN/m. Avoid use of bearing and intermediate stiffness. Use Fe $_{415}$ (E_{250}) steel.

Solution:

Maximum Moment,

$$M=\frac{wL^2}{8}=\frac{58.8\times 24^2}{8}$$

$$= 4233.6\ \text{kN}-\text{m}.$$

Maximum shear force = End reaction,

$$F = \frac{wL}{2} = \frac{58.8 \times 24}{2}$$

$$= 705.6 \text{ kN}$$

Depth of Web Plate

If stiffeners are to be avoided,

$$k = \frac{d}{t_w} \le 67$$

∴ Economical depth of web,

$$d = 3\sqrt{\frac{Mk}{f_y}} = \left(\frac{4233.6 \times 10^6 \times 67}{250}\right)^{1/3}$$

$$= 1043 \text{ mm}$$

Use 1000 mm plates,

$$t_w \ge \frac{1000}{67} \text{ i.e., } t_w \ge 14.92$$

Select t_w = 16 mm.

Thus web plate selected is 1000 mm × 16 mm.

Selection of Flange

Neglecting the moment capacity of web, area of flange required is,

$$\frac{A_f\ f_y\ d}{1.1} \ge M$$

$$\frac{A_f \times 250 \times 1000}{1.1} \ge 4233.6 \times 10^6.$$

To keep the flange in plastic category,

$$\frac{b}{t_f} \le 8.4$$

Assuming,

$$t_f = 16.8\ t_f$$

We get,

$$A_f = 16.8\, t_f\, t_f = 18628$$

$$\therefore\ t_f = 33.3 \text{ mm.}$$

Select 40 mm plates. Width of plate require = $\dfrac{18628}{40} = 465.7$

Hence use 480 mm wide and 40 mm thick plates. Section selected is shown in figure:

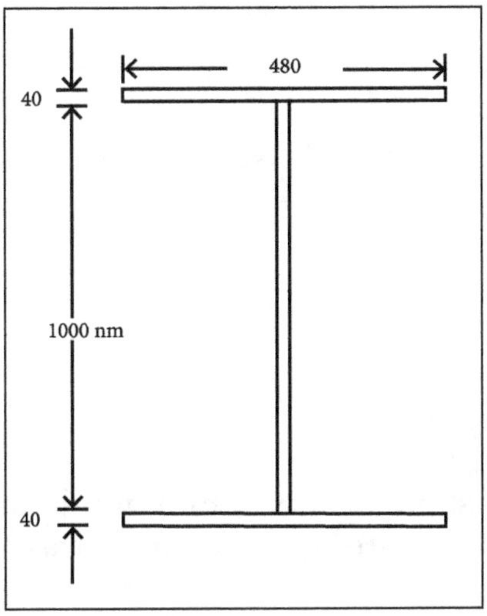

Check for the moment capacity of the girder.

Since it is assumed that only flanges resist the moment and flange is a semi-compact section.

Now,

$$I_{ZZ} = 2\left[\frac{1}{12} \times 48 \times 40^3 + 480 \times 40 \times \left(\frac{1000 + 2 \times 40}{2}\right)^2\right]$$

$$= 2 \times 5601.28 \times 106 \text{ mm}^4.$$

$$Z_C = \frac{I_{xx}}{y_{max}} = \frac{2 \times 5601.28 \times 10^6}{540} = 20.745 \times 10^6 \text{ mm}^3$$

$$M_d = \frac{20.745 \times 10^6 \times 250}{1.1} = 4714.77 \times 10^6 \text{ N} - \text{mm.}$$

$$= 4714.77 \text{ kN} - \text{m} > \text{M}.$$

Hence section is adequate.

Shear Resistance of Web

$$V_d = \frac{V_h}{\gamma_{mo}} = \frac{Vf_{yw}}{V_{mo}\sqrt{3}} = \frac{dt_w f_{yw}}{V_{mo}\sqrt{3}}$$

$$V_d = \frac{1000 \times 16 \times 250}{1.1\sqrt{3}} = 2099.455 \times 10^3 \text{ N}$$

$$= 2099.455 \text{ kN } 705.6 \text{ kN.}$$

Hence section is adequate.

No stiffeners are required.

Check for End Bearing

Bearing strength of web,

$$F_w = (b_1 + n_2)t_2 \frac{f_{yw}}{\gamma_{mo}}.$$

Assuming that the width of support is 200 mm, minimum stiff bearing provided by support = 100mm. Dispersion length n_2 = 2.5 × 40 = 100 mm.

$$F_w = (100 + 100) \times 16 \times \frac{250}{1.1}$$

$$= 727 \times 103 \text{ N} = 727 \text{ kN} > 705.6 \text{ kN}$$

Design of Weld Connecting Web Late and Flange

Maximum shear force = 705.6 kN.

Shear stress in flage at the level of junction of web and flange,

$$q = \frac{F}{bl}(\bar{a}_y)$$

$$q = \frac{705.6 \times 10^3}{480 \times 2 \times 5601.28 \times 10^6}\left[480 \times 40 \times \left(500 + \frac{40}{2}\right)\right] = 0.512 \text{ N/mm}^2$$

Shear force per mm length in junction,

$$= 0.512 \times 480 = 245.76 \text{ N}$$

The 40 mm long welds with a gap of 160 mm.

∴ Final Design = Web: 1000 × 16 mm

Flange: 400 × 40 mm

2. Let us discuss about a reverted plate girder is simply supported over an effective span of 16 m. It carries a uniformly distributed load of 80 kN/m in addition to its self-weight. And two point loads of 400 kN each at 4 m from their supports.

Solution:

Design Web and Flanges

- Plate Girder - Simply Supported.
- Effective Span = 16 m.
- udl = 80 kN/m.
- 2 Point Loads = 400 kN @ 4 m from their supports.

⇒ Computation of maximum bending Moment and Shear Force.

Factored load on span = 1.5[80 × 16 + 400 + 400] = 3120 kN,

$$= \frac{3120}{1} N/m = 3120 kN/m$$

Assuming a Self weight of total imposed load,

Self-weight of girder/m $= \frac{3120 \times 16}{200} = 249.6k$

$$\therefore \text{ Maximum B.M} = \frac{\omega l^2}{8}$$

$$= \frac{249.6 \times 16^2}{8} = 7987 \text{ kN.m}$$

$$\therefore \text{ Maximum S.F} = \frac{\omega l}{2} = \frac{249.6 \times 16}{2} = 1997 \text{ kN.}$$

Computation of Depth of Web Plate

Designing the web plate to avoid providing web stiffeners.

$$k = \frac{d}{t_w} \leq 0.67$$

$$d = \left[\frac{MK}{\delta_y}\right]^{1/3} = \left[\frac{7987 \times 10^6 \times 67}{250}\right]^{1/3} = 1288 \text{ mm}$$

Let us provide web plate of depth 1250 mm.

$\therefore$ Thickness of Web, $t_\omega \geq \frac{1250}{67} \geq 18.65\,\text{mm}.$

$\therefore$ Let us provide Web Plate of Size 1250 × 20 mm.

Computation of Flange Width

Assuming that the moment capacity of web is neglected, area of Flange required is such that,

$$\frac{A_f f_y d}{\gamma_{m_0}} \geq M$$

$$\frac{A_f \times 250 \times 1250}{1.1} \geq 7987 \times 10^6$$

$$A_f \geq 28114\ \text{mm}^2.$$

Let us provide the flange to be under the semi compact category,

$$\frac{b}{t_f} \leq 13.6 t_f = \frac{b}{12}$$

$$A_f = b_f t_f$$

$$= (12\,t_f)\,t_f$$

$$b_f = \sqrt{\frac{28114}{12}}$$

$$t_f = 48.4\ \text{mm}$$

$$t_f = 50\ \text{mm}.$$

$$b = 12 \times 50$$

$$= 600\ \text{mm}.$$

Hence Let us use 600 × 50 mm plates for the flanges and 1250 × 20 mm plates for the web.

3. Let us design a riveted plate girder using Fe 415 steel span 22 m to carry a load 25 kN/m.

Solution:

Permissible bending stress = 165 N/mm²

Shear stress = 100 N/mm²

Total super imposed to girder = 22 × 25

$$= 550 \text{ kN}$$

$$\text{Self-Wet of girder } = \frac{\text{total super imposed} \times \text{span in m load}}{300}$$

$$= \frac{550 \times 22}{300}$$

$$= 40.33 \text{ kN.}$$

Total load = 550 + 41 = 591 kN.

$$S.F = R_A = R_B = \frac{\text{Total load}}{2}$$

$$= \frac{591}{2} = 295 \text{ say } 296 \text{ kN}$$

$$B.M = \frac{591 \times 22}{8} = 1625.25 \text{ kN.}$$

Design of the Web

Economic effective depth,

$$= 5.5\ 3\sqrt{\frac{M}{f}}$$

$$= 5.5\ 3\sqrt{\frac{1625.25 \times 10^6}{165}}$$

$$= 1178.98 \approx 1180 \text{ mm.}$$

Actual depth = 1180 – 2(5) = 1170 mm

$$h_c = h - 2h_f$$

$$= 1180 - 2(150) = 880 \text{ mm}$$

$$\text{Area of Web } = \frac{s}{f_s}$$

$$F = 296 \text{ kN} = \frac{296\,(1000)}{100}$$

$$= 2960 \text{ mm}^2.$$

$\therefore$ Thickness of Web = $\dfrac{\text{Area of Web}}{\text{Clear of Web}}$

Ratio of Clear Web $= \dfrac{2960}{880} = 3.36$

$$= \frac{880}{4} = 220$$

Provide 5 mm,

$$= \frac{880}{5} = 176 < 200$$

$\therefore$ Provide Web plate 1180 mm × 5 mm

Design of Flange

$$A_f = \frac{M}{fh} - \frac{1}{8} A_w$$

$$= \frac{1625.25 \times 10^6}{(165)(1180)} - \frac{1}{8}(1180 \times 5)$$

$$= 7609.95 \approx 7610 \text{ mm}^2.$$

Provide 150 mm × 150 mm × 15 flange.

Gross area of two angle = 8566 mm²

20 + 1.5 = 21.5mm

Net area provide,

$$= 8566 - 4(21.5)(15)$$

$$= 7266 \text{ mm}^2$$

$$= 7610 - 7266 = 344 \text{ mm}^2$$

Net width of cover plate,

$$= 320 - 2(21.5)$$

$$= 277 \text{ mm}$$

Thickness of cover plate = $\dfrac{344}{277}$ = 1.24 mm.

$\therefore$ The size of each flange = 320 × 2.5 mm.

4.5 Intermediate and Bearing Stiffeners

Purpose of Providing Bearing Stiffeners

The function of a bearing stiffness is to preclude any crushing of the web at locations of heavy concentrated loads. Thus bearing stiffener transfer heavy reactions or concentrated load to the full depth of the web. They are placed in pairs on the web of plate girders at unframed girder ends and where required for concentrated load.

Formula for calculating the thickness of beam bearing plate,

$$t=\sqrt{\frac{3w_1 a^2}{\sigma_{bs}}}$$

Where,

w_1 = Unit pressure from masonry,

σ_{bs} = Bending stress in slab base (185 MPa),

t = Thickness of bearing plate,

a = Length of cantilever portion of bearing plate.

Problems

1. Let us discuss the problem based on a bearing stiffener for a welded-plate girder with the following specifications: Web = 1000 mm × 6 mm, Flange = 2 Nos. of 320 × 20 mm plate on each side Support reaction = 300 kN and width of support = 300 mm.

Solution:

Outstand of stiffener,

= 0.45 × Width of Flange

= 0.45 × 300 = 135 mm

Bearing area required= $\frac{600 \times 10^3}{180}$ = 3,333.3 mm2

Thickness of bearing stiffness plate,

$$= \frac{3{,}333.3}{2\times(135-10)} = 13.33 \text{ mm say 15mm.}$$

Total area of the stiffener,

= 2(15 × 135) + (120 × 6) = 4,770 mm².

Induced compressive stress $=\frac{600\times10^3}{4{,}770}$ = 125.8 N/mm2.

Moment of inertia of the stiffener,

$$I=\frac{15\times276^3}{12}+\frac{(120-15)6^3}{12}$$

$$I=2{,}628.261\times10^4\,mm^4$$

$$\gamma=\sqrt{\frac{2625.261\times10^4}{4770}}=74.22\text{ mm.}$$

Slenderness ratio $=\frac{l}{r}=\frac{0.7\times1000}{74.22}=9.43.$

Safe compressive stress corresponding to the slenderness ratio 150 N/mm².

Connections vertical load transmitted to the stiffener per mm height of the stiffener =

$q_1=\frac{600\times10^3}{1000}$ = 600 N/mm.

Provide 5 mm wld, strength of the weld per mm length,

S_ω = 0.7 × 5 × 110 = 385 N/mm

l_ω = 50 mm

Spacing of the weld units,

$$=\frac{4S_\omega l_\omega}{q_1}$$

$$=\frac{4\times385\times50}{600}$$

= 128 mm.

Clear spacing between weld units,

= 128 – 50 = 78mm.

Maximum permissible clear spacing between weld units,

= 12t = 12 – 6 = 72mm.

Provide a clear spacing of 70 mm.

4.6 Flange and Web Splices

Web Splice

Web splice plates and their connections shall be designed for shear, the moment due to the eccentricity of the shear at the point of splice and the portion of the flexural moment assumed to be resisted by the web at the point of the splice.

Flange Splices

A joint in the flange element provided to increase the length of flange plate is known as flange splice. The flange splices should be avoided as far as possible. Generally the flange plates can be obtained for full length of the plate girder. In spite of the availability of full length of flange plates sometimes it becomes necessary to make flange splices. Flange joints should not be located at the points of maximum bending moment. >

Problems

A plate girder span 15 m is made up of web plate 1600 mm × 8 mm × flange angle 150 mm × 115 mm × 10 mm flange plate 480 mm × 10 mm. It carries a udl load 100 kN/m including its own wet. Let us design and sketch web splice at 5 m to 10 m end.

Solution:

Load on the girder = 100 kN

Span = 15 m

Reaction support,

$$= \frac{15(100)}{2}$$

= 750 kN

= S = 750 – 100(5)

= 250 kN.

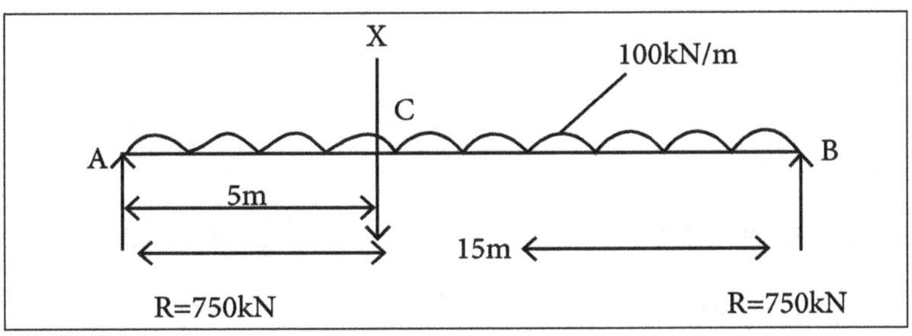

B.M Splice = 750 (5) – 100 (5) $\left(\frac{5}{2}\right)$

Let 20 mm = 2500 kN.m

Diameter of rivet,

= 20 + 1.5

= 21.5 mm.

Net Area of One Flange

= 2(2552) – 4(21.5) (10)

= 4244 mm²

= 2(480 – 2 × 21.5) (10)

= 8740 mm²

Net Area of flange,

$A_f = 4244 + 8740$

$= 12984 \text{ mm}^2$

One-Eight Web,

$$=\frac{1}{8}A_W$$

$$=\frac{1}{8}(1600 \times 8) = 1600 \text{ mm}^2$$

$$M_w = \left[\frac{\frac{1}{8}A_w}{A_f + \frac{1}{8}A_w}\right] M$$

$$= \left[\frac{1600}{12984 + 1600}\right] 2500$$

$M_w = 274.273 \text{ kN}\cdot\text{m}.$

Design of Web Splice

= 1600 – 2(115) = 1370

= 1370 – 2 × 5 = 1360

$$\Rightarrow \frac{1360}{60} = 22.678 \text{ say22 rivets}$$

$$\therefore \text{Vertical distance } V = \frac{\text{Shear force}}{\text{No. of rivets}}$$

$$= \frac{250 \times 10^3}{3 \times 22} = 3788 \text{M}$$

Horizontal resistance $= k \cdot y$

$$k = \frac{M_w}{\sum y^2}$$

$$\sum y^2 = 3\{630^2 + 570^2 + 510^2 + 450^2 + 390^2 + 330^2 + 270^2 + 210^2$$

$$+150^2 + 90^2 + 30^2\} \text{ mm}^2$$

$$= 3(1593900)$$

$$= 4781700 \text{ mm}^2$$

$$k = \frac{274.273 \times 10^6}{4781700}$$

$$= 57.36 \text{ N/mm.}$$

$$H = k \cdot y_{max} = 57.36\ (630)$$

$$= 3613.7 \text{ N}$$

Resultant resistance,

$$= \sqrt{V^2 + M^2}$$
$$= \sqrt{3788^2} + 36137^2$$
$$= 36335 \text{ N}$$

$$= 2\left(\frac{Td^2}{4}\right) f_s$$

$$= 2\left(\frac{3.14 \times 21.5^2}{4}\right) 100$$

$= 72610 \text{ N} = dt\, f_b$

$= 21.5 \times 2 \times 300$

$= 51600$

∴ Safe load = least value = 51600 N 36335 N.

Thickness of Splice Plate:

$$t_o = \frac{3 M_W h}{f\left(n_o^3 - 12d \sum y^2\right)}$$

= 1000 + 2

= 1600 + 2(5) = 1610 mm

d = 21.5 mm

h_o = 1360 mm

$$\sum y^2 = y_1^2 + y_2^2 + y_3^2 \; ....$$

$$= \frac{4781700}{3} = 1593900 \text{ mm}^2$$

$$f = \frac{\text{B.M at set 10 n}}{n\left(A_f + \frac{1}{8} A_w\right)}$$

$$= \frac{2500 \times 10^6}{1610\left(12984 + \frac{1}{8}(1600 \times 8)\right)} = 106.47 \text{ N/mm}^2$$

$$t_o = \frac{3 \times 274.273 \times 10^6 \times 1610}{106.47\left[13603 - 12(21.5)(1593900)\right]}$$

t = 6 mm

Width of Splice,

= 40 + 60 + 60 + 40 + 10 + 10 + 60 + 60 + 40

= 410 mm.

The girder shown in Figure is fully restrained against lateral buckling throughout its span. The span is 36 m and carries two concentrated loads as show in Figure. Design a plate girder:

- Yield stress of steel, $fy = 250\ N/mm^2$.
- Material factor for steel, $\gamma m = 1.15$.
- Dead Load factor, $\gamma fd = 1.50$.
- Imposed load factor, $\gamma fl = 1.50$.

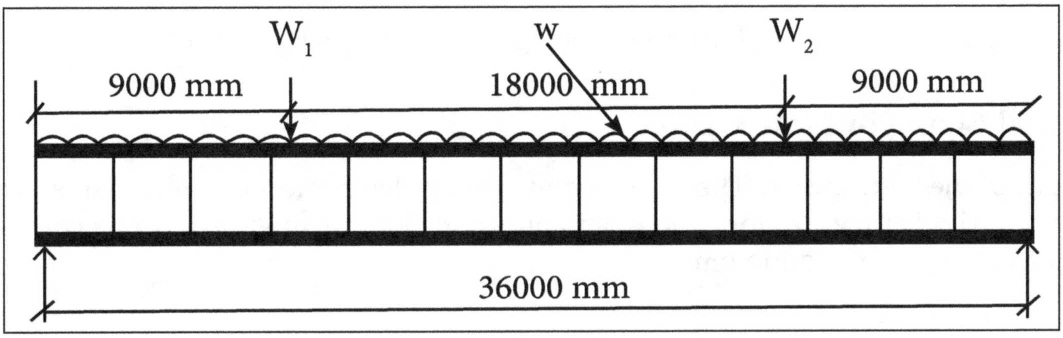

Loading

Dead load,

Uniformly distributed load, $w_d = 18\ kN/m$

Concentrated load, $W_1d = 180\ KN$

Concentrated load, $W_2d = 180\ KN$

Live load,

Uniformly distributed load, $W = 35\ kN/m$

Concentrated load, $W_1l = 400\ kN$

Concentrated load, $W_2l = 400\ kN$

Factored Loads $w' = w_d * \gamma fd + wl * \gamma fl = 18 * 1.5 + 35 * 1.5 = 79.5\ kN/m$

$W'_1 = W_{1d} * \gamma fd + W_{1l} * \gamma fl = 180 * 1.5 + 400 * 1.5 = 870\ kN$

$W'_2 = W_{2d} * \gamma fd + W_{2l} * \gamma fl = 180 * 1.5 + 400 * 1.5 = 870\ kN$

Bending Moment and Shear Force

	Bending moment (kN-m)	Shear force (kN)
UDL effect	$\frac{w}{8}=\frac{79.5\times 36\times 36}{8}=12879$	$\frac{w\ell}{2}=1431$
Concentrated load effect	$\frac{w\ell}{4}=870\times 9=7830$	w = 870
Total	20709	2301

The design shear forces and bending moments are shown in figure (a).

Initial Sizing of Plate Girder

Depth of the plate girder: The recommended span/depth ratio for simply supported girder varies between 12 for short span and wo for long span girder. Let us consider depth of the girder as 2600 mm.

$$\frac{\ell}{d}=\frac{36000}{2600}=138$$

Depth of 2600 mm is acceptable. (For drawing the bending moment and shear force diagrams, factored loads are considered):

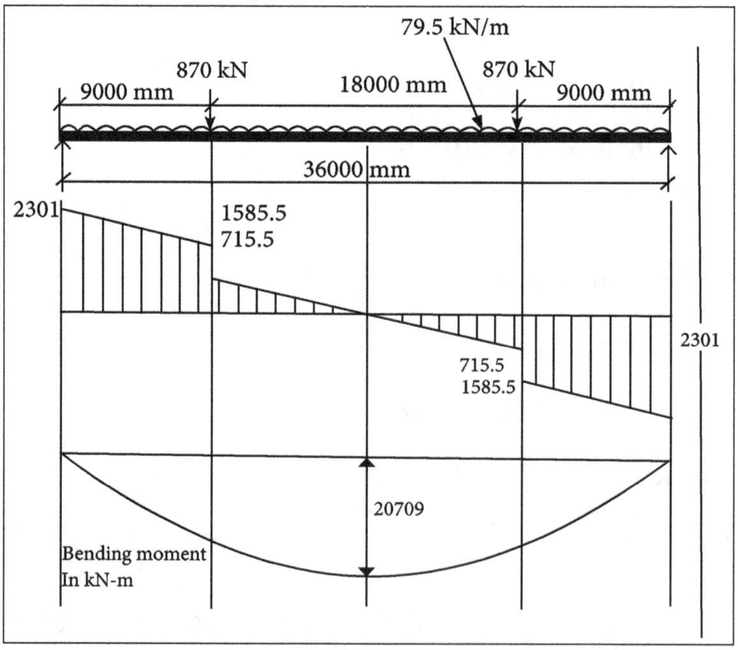

(a) Bending moment and shear force diagram.

Flange,

$$Py = 250/1.15 = 217.4\ N/mm^2$$

Single flange area,

$$A_f = \frac{M_{max}}{d_{py}} = \frac{20709 \times 10^6}{2600 \times 217.4} = 366375.5 mm^2$$

By thumb rule, the flange width is assumed as 0.3 times the depth of the section.

Try 780 X 50 mm, giving an area = 39000 mm².

Web: Minimum web thickness for plate girder in buildings usually varies between 10 mm to 20 mm. Here, thickness is assumed as 16mm. Hence, web size is 2600 X 16 mm.

Section Classification Flange

$$s = \left\{\frac{250}{f_y}\right\}^{\frac{1}{2}} = \left\{\frac{250}{250}\right\}^{\frac{1}{2}} = 1.0$$

$$b = \frac{B - t}{2} = \frac{780 - 16}{2} = 382$$

$$\frac{b}{T} = \frac{382}{50} = 7.64 < 9_g$$

Hence, Flange is compact section.

Web

$$\frac{d}{t} = \frac{2600}{250} = 162.5 > 67_g$$

Hence, the web is checked for shear buckling.

Checks

Check for serviceability:

$$\frac{d}{250} = \frac{2600}{250} = 10.4\ mm < 1$$

Since,

$$t > \frac{d}{250}$$

Web is adequate for serviceability.

Check for Flange Buckling in to Web

Assuming stiffener spacing, a > 1.5 d

$$t \geq \frac{d}{294}\left(\frac{\quad}{250}\right)^{-} = \frac{2600}{294} \times \left(\frac{217.4}{250}\right)^{-} = 8.2\,\text{mm}$$

Since, t (=16 mm) > 8.2 mm, the web is adequate to avoid flange buckling into the web.

Check for moment carrying capacity of the flanges:

The moment is assumed to be resisted by flanges alone and the web resists shear only.

Distance between centroid fo flanges, hs = d + T = 2600 + 50 = 2650 mm

$Af = B * T = 780 * 50 = 39000\ mm^2$

$Mc = Pyf * Af * hs = 217.4 * 39000 * 2650 * 10\text{-}6 = 222468.3$ kN-m

> 20709 kN-m

Hence, the section in adequate for carrying moment and web is designed for shear.

Web Design

The stiffeners are spaced as shown in figure named 'Effective section'. Three different spacing values 2500, 3250 and 3600 mm are adopted for trail as shown in figure.

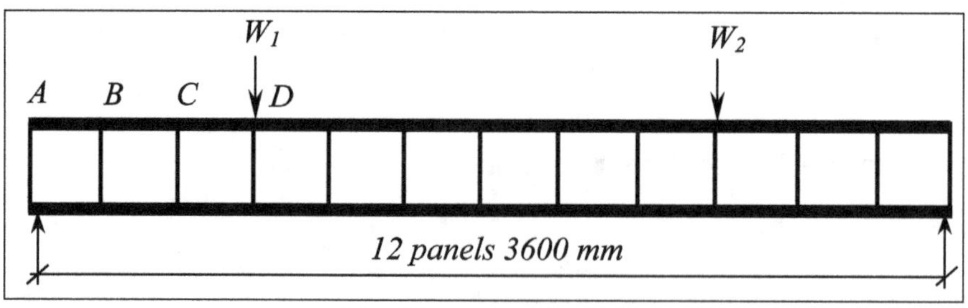

Trial stiffener arrangement.

End panel (AB) design,

d = 2600 mm t = 16 mm.

Maximum shear stress in the panel is,

$$f_v = \frac{F_{VA}}{dt} = \frac{2301*10^3}{2600*16} = 55.3 \text{ N/mm}^2$$

$$\frac{a}{d} = \frac{2500}{2600} = 0.96$$

$$\frac{d}{t} = \frac{2600}{16} = 1625$$

Calculation of critical stress,

$$q_e = (\text{when } a/d \leq 1) = \left[0.75 + 1/(a/d)^2\right]\left[1000/(d/t)\right]^2$$

$$= \left[0.75 + 1/(0.96)^2\right]\left[1000/(165.5)\right]^2$$

$$= 69.5 \text{ N/mm}^2$$

Slenderness parameter,

$$\lambda_y = \left[0.6\left(f_{yw}/\gamma_m\right)/qc\right]^{\frac{1}{2}}$$

$$= \left[0.6\,(250/1.15)69.5\right]^{\frac{1}{2}}$$

$$= 1.37 > 1.25$$

Hence, Critical shear strength ($q_{cr} = q_e$) = 69.5 N/mm²

Since,

$$f_v < qc_r\ (55.3 < 69.5)$$

Tension field action need not be utilized for design.

Checks for the End Panel AB

End panel AB should also be checked as a beam (Spanning between the flanges of the girder) capable of resisting a shear force R_{tf} and a moment M_{tf} due to anchor forces. (In the following calculations boundary stiffeners are omitted for simplicity).

Check for shear capacity of the end panel,

$$H_q = 0.75\,dt\ P_y \left[1 - \frac{q_{cr}}{0.6 P_y}\right]^{\frac{1}{2}}$$

$$q_{cr} = 69.5\ N/mm^2$$

$$H_q = 0.75 * 2600 * 16 * 250/1.15\left[1 - \frac{69.5}{0.6*(250/1.15)}\right]^{\frac{1}{2}} - 4636\,kN$$

$$R_g = \frac{H_q}{2} = \frac{4636}{2} = 2318\,kN$$

$$A_v = t.a = 16 * 2500 = 4000\,mm^2$$

$$P_v = 0.6P_{yw}A_{\frac{1}{2}} = 0.6*(250/1.15)*40000/1000 = 5217\ kN$$

Since,

$R_{tf} < P_v$, the end panel can carry the shear force.

Check for moment capacity of end panel AB,

$$M_{tf} = \frac{H_q d}{10}$$

$$= \frac{4636*2600}{10}*10^{-3} = 1205\,kN-m$$

$$y = \frac{a}{2} = \frac{2500}{2} = 1250$$

$$I = \frac{1}{12}ta^3 = \frac{1}{12}*16*2500^3 = 2083\times10^7\ mm^4$$

$$M_q = \frac{1}{y}py = \frac{2083*10^3}{1250}*(250/1.15)*10^{-6} = 3623\,kN-m$$

Since,

$M_{tf} < M_q$ $\quad$ $(1205<3623)$

The end panel can carry the bending moment.

Design of Panel BC

Panel BC will be designed using tension field action,

d = 2600 mm; t = 16 mm

$$f_y = \frac{F_{VB}}{dt} = \frac{2102.3*10^3}{2600*16} = 50.5 N/mm^2$$

$$\frac{a}{d} = \frac{3250}{2600} = 1.25$$

$$\frac{d}{t} = \frac{2600}{16} = 162.5$$

$$P_y = \frac{250}{1.15} = 217.4 N/mm^2$$

Calculation of basic shear strength, qb:

Elastic critical stress, q_s (when a/d >1),

$$= \left[1.0 + 0.75/(a/d)^2\right]\left[1000/(d/t)\right]^2$$

$$= \left[1.0 + 0.75/(1.25)^2\right]\left[1000/(162.5)\right]^2$$

$$= 56.0 \text{ N}/mm^2$$

Slenderness parameter, λ_n,

$$= \left[0.6(f_{yw}/\gamma_m)/q_e\right]^{\frac{1}{2}}$$

$$= \left[0.6(250/1.15)/56.0\right]^{\frac{1}{2}}$$

$$= 2.33 > 125$$

$$\text{Hence, Critical shear stength} = 56.0 \text{ N}/mm^2 (q_{cr} = q_s)$$

$$\phi_1 = \frac{1.5 q_{cr}}{\sqrt{1 + \left(\frac{a}{d}\right)^2}} = \frac{1.5*56.0}{\sqrt{1+(1.25)^2}} = 52.5$$

$$y_b = \left(P_{yw}{}^2 - 3_{q_{cr}}{}^2 - \phi_1{}^2\right)^{\frac{1}{2}} - \phi_1 = \left(217.4^2 - 3*56.0^2 + 52.5^2\right)^{\frac{1}{2}} - 52.5 = 149$$

$$q_b = q_{cr} + \frac{y_b}{2\left[\frac{a}{c} + \sqrt{1+\left(\frac{a}{d}\right)^2}\right]} = 56.0 + \frac{149}{2\left[125 + \sqrt{1+(125)^2}\right]} = 82.1 N/mm^2$$

Since,

$$q_b > f_s \quad (82.1 > 50.5)$$

Panel BC is safe against shear buckling.

Design of Stiffeners

- Load bearing stiffener at A.
- Design should be made for compression force due to bearing and moment.
- Design force due to bearing, Fb = 2301 kN.

Force (Fm) due to moment M_{tf}, is,

$$F_m = \frac{M_{tf}}{a} = \frac{1205}{2500} * 10^3 = 482\,kN$$

Total compression = $F_c = F_b + F_m$ = 2301 + 482 = 2783 kN

Area of Stiffener in contact with the flange, A.

Area (A) should be greater than,

$$= \frac{0.8\,F_c}{P_{ys}}$$

$$\frac{0.8F_c}{P_y} = \frac{0.8 * 2783}{217.4} * 10^3 - 10241\ mm^2$$

Try stiffener of 2 flats of size 270 x 25 mm thick.

Allow 15 mm to cope for web/flange weld,

$$A = 255 * 25 * 2 = 12750\ mm^2 > 10241\ mm^2$$

Bearing check is OK.

Check for Outstand

Outstand from face of web should not be greater than,

$$= 20\ t_y\ s$$

$$s = \left\{\frac{250}{f_y}\right\}^{\frac{1}{2}} = \left\{\frac{250}{250}\right\}^{\frac{1}{2}} = 10$$

$$\text{Outstand } b_s = 250\,mm < 20t_g s(= 20 * 25 * 1.0 = 500)$$

$$b_s = 250\,mm < 13.7\,t_g\,s(= 13.7 * 25 * 1.0 = 342.5)$$

Hence, outstand criteria is satisfied.

Check Stiffener for Buckling

The effective stiffener section is shown in figure:

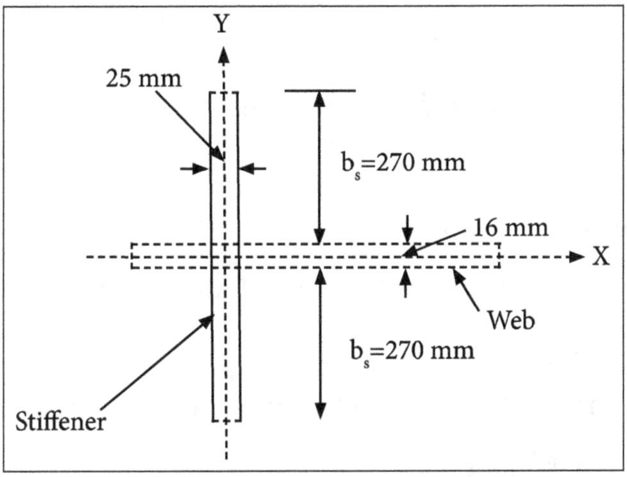

End bearing stiffener.

The buckling resistance due to web is neglected here for the sake of simplicity.

$$I_x = \frac{25 \times 556^3}{12} - \frac{1}{12} \times 25 \times 16^3 = 35807 \times 10^4 \text{ mm}^4$$

$$A_e = \text{Effective area} = 270 \times 25 \times 2 = 13500 \text{ mm}^2$$

$$r_x = \left[\frac{I_x}{A_e}\right]^{\frac{1}{2}} - \left[\frac{35807 \times 10^4}{13500}\right]^{\frac{1}{2}} = 162.8 \text{ mm}$$

Flange is restrained against rotation in the plane of stiffener,

Then,

$I_e = 0.71 = 0.7 * 2600 = 1820$

$$\lambda = \frac{I_e}{r} = \frac{1820}{162.8} = 11.2$$

For $f_y = 250 \text{ N/mm}^2$ and $\lambda = 11.2$

$$\sigma_c = 250 \text{ N/mm}^2$$

Buckling resistance of stiffener is,

$$P_c = \sigma_c A_e / \gamma_m = 250 * 1350 / 1.15 = 2935 \text{ kN}$$

Since,

$$F_c < P_c\,(2783 < 2935)$$

Therefore, stiffener provided is safe against buckling.

Check Stiffener A as a Bearing Stiffener

Local capacity of the web.

Assume, stiff bearing length $b_1 = 0$,

$n_2 = 2.5 * 50 * 2 = 250$ BS 5950 : Part -1, Clause 4.5.3

$$P_{crip} = (b_1 + n_2)\, tP_{yw}$$

$$= (0 + 250) * 16 * (250/1.15) * 10\text{-}3 = 870 \text{ kN}$$

Bearing stiffener is designed for F_A,

$$F_A = F_x = P_{crip} = 2783 - 870 = 1931 \text{ kN}$$

Bearing capacity of stiffener alone,

$$P_A = P_{ys} * A = (50/1.15) * 13500/1000 = 2935 \text{ kN}$$

Since,

$$F_A < P_A\,(1931 < 2935)$$

The designed stiffener is OK in bearing.

Stiffener A - Adopt 2 flats 270 mm X 25 mm thick.

Design of Intermediate Stiffener at B

Stiffener at B is the most critical one and will be chosen for the design.

Minimum Stiffness,

$$I_e \geq 0.75\, dt^3 \quad \text{for } a \geq d\sqrt{2}$$

$$I_e \geq \frac{0.75\, dt^3}{a^3} \quad \text{for } a < d\sqrt{2}$$

$$d\sqrt{2} = \sqrt{2} \times 2600 = 3677 \text{ mm}$$

$$a < d\sqrt{2} \qquad (3250 < 3677)$$

Conservatively ‘t’ is taken as actual web thickness and minimum ‘a’ is used.

$$\frac{1.5\ d^3\ t^3}{a^2} = \frac{15 \times 2600^3 \times 16^3}{3250^2} = 1022 - 10^4\ mm^4$$

Try intermediate stiffener of 2 flats 120 mm X 14 mm.

Check for Outstand

Outstand of the stiffener $\leq 13.7\,t_g s$

$13.7\,t_g s = 13.7 \times 14 \times 1.0 = 192\,mm$

Outstand = 120 mm (120 < 192),

Hence, outstand criteria is satisfied.

Buckling Check

Stiffener force, $F_q = V - V_s$

Where,

V = Total shear force

$V_s = V_{cr}$ of the web

$a / d = 3600 / 2600 = 1.38$

$d / t = 2600 / 16 = 162.5$

Elastic critical stress, q_s (when a/d > 1),

$$= \left[1.0 + 0.75/(a/d)^2\right]\left[1000/(d/t)\right]^2$$

$$= \left[1.0 + 0.75f(1.38)^2\right]\left[1000/(162.5)\right]^2$$

$$= 52.8\ N/mm^2$$

Slendemess parameter, λ_n

$$= \left[0.6\left(f_{yw}/\gamma_m\right)/qe\right]^{\frac{1}{2}}$$

$$= \left[0.6(250/1.15)/52.8\right]^{\frac{1}{2}}$$

$$= 2.47 > 1.25$$

Hence, Critical shear strength,

$= 52.8 \text{ N/mm}^2 \ (q_{cr} = q_e)$

$V_{cr} = q_{cr}dt = 52.8 * 2600 * 16 * 10\text{-}3 = 2196 \text{ kN}$

Buckling resistance of intermediate stiffener at B:

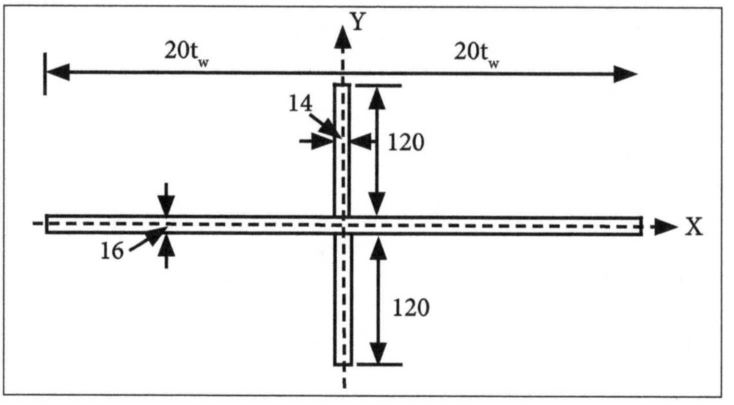

Effective section.

$$20t_w = 20 \times 16 = 3230 \text{ mm}$$

$$I_e = \frac{1}{12} \times 14 \times 256^3 + \frac{640 \times 16^3}{12} - \frac{14 \times 16^3}{12} = 1979 \times 10^4 \text{ mm}^4$$

$$A = 240 \times 14 + 640 \times 16 = 13600 \text{ mm}^2$$

$$r_x = \left[\frac{1979 \times 10^4}{13600}\right]^{\frac{1}{2}} = 38.1$$

$$I_e = 0.7 \times 2600 = 1820$$

$$\lambda = \frac{I_e}{r_x} = \frac{1820}{38.1} = 48.0$$

$$\text{For } f_y = 250 \text{N/mm}^2 \text{ and } \lambda = 48.0$$

From table of axially compressed columns,

$\sigma_c = 213.2 \text{ N/mm}^2$

Buckling resistance = (213.2/1.15) * 13600 * 10-3 = 2521 kN

Shear force at B, VB = 2301-{(2301 - 1585.5)*(2500/9000)] = 2102 kN

Stiffener force, Fq = [21201 - 2196] < 0

And,

Fq < Buckling resistance

Hence, intermediate stiffener is adequate.

Intermediate stiffener at B - Adopt 2 flats 120 mm X 14 mm.

Intermediate stiffener at E (Stiffener subjected to external load):

Minimum stiffness calculation,

$$a = 3600 \qquad d\sqrt{2} = 3677$$

$$a < d\sqrt{2} \qquad (3600 < 3677)$$

$$I_e \geq \frac{1.5 d^3 t^3}{a^2} - \frac{1.5 \times 2600^3 \times 16^3}{3600^2} = 833 \times 10^4$$

Try intermediate stiffener 2 flats 100 mm X 12 mm thick,

$$(I_s)_{provided} = 1007 \times 10^4 \text{ mm}^4$$

$$(I_s)_{provided} > I_s \qquad \left[1007 \times 10^4 > 833 \times 10^4\right]$$

Hence, OK.

Buckling Check:

$$\frac{F_q - F_x}{P_q} + \frac{F_x}{P_x} + \frac{M_s}{M_{ys}} \leq 1$$

$$F_q = V - V_s \qquad V = 1585.5 \text{kN}$$

$$V_s = V_{cr} = q_{cr}\, dt = 52.8 \times 2600 \times 16 \times 10^{-3} = 2196 \text{kN}$$

F_q is negative and $F_q - F_x = 0$

$$M_s = 0$$

$$F_x = 870 \text{kN}$$

Buckling resistance of load carrying stiffener at D:

(Calculation is similar to stiffener at B),

$$20 t_w = 20 \times 16 = 320 \text{mm}$$

$$I_x - \frac{1}{12}\times 12\times 216^3 + \frac{640\times 16^3}{12} - \frac{12\times 16^3}{12} - 1029\times 10^4\ mm^4$$

$$A = 200\times 12 + 640\times 16 = 12640\,mm^2$$

$$r_x = \left[\frac{1029\times 10^4}{12640}\right]^{\frac{1}{2}} = 285$$

$$I_e = 0.7\times 2600 - 1820$$

$$\lambda = \frac{I_e}{r_x} = \frac{1820}{28.5} = 63.9$$

For $f_y - 250\,N/mm^2$ and $\lambda = 63.9$

From table of axially compressed columns,

σ_c = 180 N/mm²

Buckling resistance, P_x (180/1.15) * 2640 * 10-3 = 1978 kN.

F_x/P_x = 870/1978 = 0.44 <1.0

Hence, Stiffener at D is OK against buckling.

Stiffener at D - Adopt flats 100 mm X 12 mm thick.

Web check between stiffeners: fed = w1 / t = 79.5 / 16 = 4.97 N/mm²

When compression flange is restrained against rotation relative to the web.

$$P_{ed} = \left[2.75 + \frac{2}{(a/d)^2}\right]\frac{E}{(d/t)^2} = \left[2.75 + \frac{2}{\left(\frac{3600}{2600}\right)^2}\right]\frac{200000}{\left(\frac{2600}{16}\right)^2}$$

$$= \frac{3.79\times 20000}{26406} = 28.7\ N/mm^2$$

Since,

$$f_{ed} \leq P_{ed}$$

[4.97 <28.7], the web is OK for all panels.

Final Girder

(All dimensions are in mm).

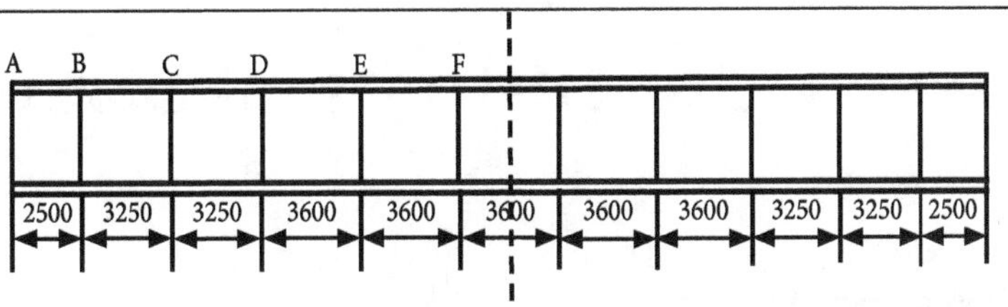

(a) Longitundinal section of plate girder

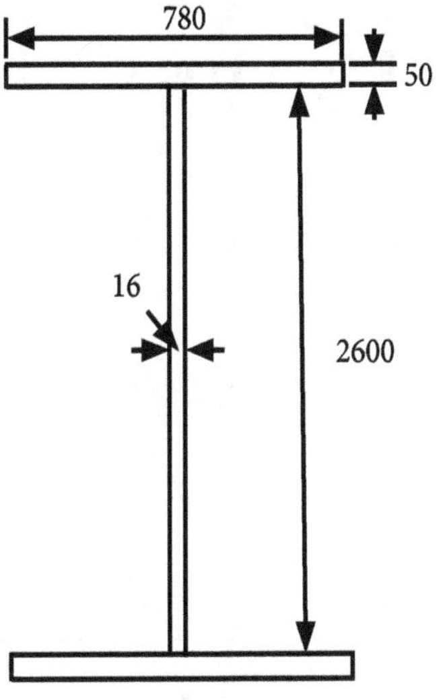

(b) Flange and web proportions

Final girder.

5

Roof Trusses and Industrial Structures

5.1 Roof Trusses

Different Components of Roof Truss

- Top chord members.
- Bottom chord members.
- Struts.
- Slings.
- Sag Tie.

Sag Rod

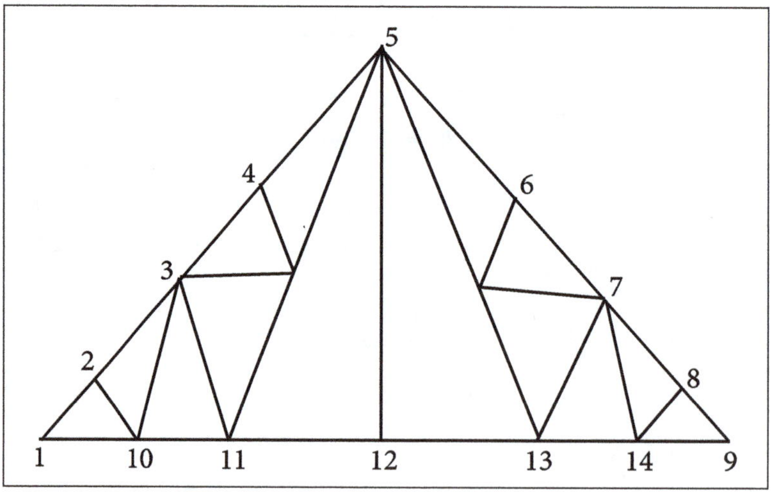

In the roof truss shown in Figure, member 5-12 is called the sag rod. If no-load is acting at joint 12; the member is not subjected to any force. Even then this member is provided to reduce the sag in the member 11-13 and hence it is called as sag rod.

General Guidelines for Fixing Spacing of Roof Stresses

- 3 to 4.5 m up to 15 m span.

- 4.5 to 6.0 m up to 15 – 30 m span.
- For span more than 40 m, spacing cf 12 - 15 m may be used.

Basis is Live Load Considered in the Design of Roof Truss

Live Load for design of sheets and purling:

- Live Load upto 10° Slope = 0.75 N/m^2.
- For more than 10° Slope = 0.75 – 0.02 [θ – 10] where θ is Slope of Sheeting.
- However a minimum of 0.4 kN/m^2 as Live Load should be considered in any case.
- For the design of trusses the above L.L may be reduced to 2/3 rds.

The Role of End Bearing in Roof Trusses

End bearing are provided at the supports of roof trusses in the form of steel plate anchored on the column in the form of anchor bolts in order to transfer the reaction from the truss member onto the columns.

Problems

1. A roof terms shed is to be build in Lucknow for an industry. The six of shed is 24 m × 40 m. The height of building is 12 m at the ever. Let us determine the basic wind pressure.

Solution:

- From wind zone map of country V_b = 47 m/sec.
- Risk co-efficient k_1 = 1.0.
- Terrain, height and structure size factor k_2,

  k_2 = 0.88 if h = 10 m

  = 0.34 if h = 15m

For h = 12 m, k2 = $0.88 + (0.94 - 0.88)\frac{2}{5} = 0.904$

Topography factor k_3,

$$k_3 = 1 + C_s$$

$$c = \frac{2}{\varepsilon} = 0$$

k = 1.0.

Design Wind Speed

$V_2 = k_1k_2k_3V_b$

$V_2 = 1.0 \times 0.904 \times 1.0 \times 47$

$V_2 = 42.488$ m/sec.

Basic Wind Pressure

$PZ = 0.6\ V_z^2 = 0.6 \times 42.4882.$

$P_z = 1083 N/m^2.$

$P = 1.083\ kN/m^2.$

5.2 Roof and Side Coverings

A roof of a building envelope both the covering on the uppermost part of a building or shelter which provides protection from animals and weather notably rain, but also heat, wind and sunlight and the framing or structure which supports the covering:

- All the materials laid on the roof frame; includes sheathing, the outer cladding materials, asphalt paper, etc.
- A roof covering 1 which is not readily flammable and does not slip from position. The following classes have these and additional properties.
- Class A is effective against severe fire exposure, does not carry or communicate fire and affords a fairly high degree of fire protection to the roof-deck.
- Class B is effective against moderate fire exposure does not readily carry or communicate fire and affords a moderate degree of fire protection to the roof-deck.
- Class C is effective against light fire exposure does not readily carry or communicate fire and affords a slight degree of fire protection to the roof-deck.

5.3 Design of Purlin, Elements of Truss and End Bearing

Let us calculate the wind load while designing roof trusses.

Solution:

Wind Load While Design Roof Trusses

- Determine Basic Wind Speed [Vb].

Refer IS875 Part 3 - CI.5.2.

- Design Wind Speed [Vz],

  $V_z = K_1 K_2 K_3 V_b$

Where,

$K_1 \rightarrow$ risk Co-efficient [Table - IS875 - Part 3]

$K_2 \rightarrow$ terrain, height and structure size factors

[Table - IS875 - Part 3]

$K_3 \rightarrow$ topography factor [C1.5.4 of IS875 - Part - 3].

- Design Wind Pressure [Pz],

  $P_Z = 0.6 V_Z^2$

- Design Wind Pressure on Roof [F],

Refer IS875 - Part 3 of Cl.6.2.1

$F = [C_{pe} - C_{\pi}] A P_d.$

1. Let us determine the driven loads on the pudine of an industrial building near Visakhapatnam given: Class of building: General with life of 50 years, Terrain: Category2, Maximum dimension: 40 m, Width of building: 15m, Height at use level: 8 m, Topography: θ less than 3°.

Solution:

Permeability: Medium

Span of truss: 15m

Pitch: $\frac{1}{5}$

Sheeting: A.C. sheets

Spacing of purlins: 1.35 m.

Spacing of trusses: 4 m.

1. D.L. Calculations

Weight of sheeting including laps and connectors = 170 N/m².

Self-weight of purlin (assumed) = 100N/m².

Total D.L. on purlin,

$= 170 + 100$

$= 270\ N/m^2.$

Spacing of purlins = 1.35m.

D.L. on purlin = 270 × 1.35 = 364.5 N/m.

2. Live Load

Span of truss = 15 m

$$\text{Pitch} = \frac{1}{5}\ \text{Rise} = \frac{1}{5} \times 15 = 3\ \text{m}.$$

$$\tan\theta = \frac{3}{7\cdot 5}\ \text{or}\ \theta = 621.8°$$

Live load on purlin = 750 – (21.8 – 10) × 20 = 514 N/m².

Spacing of purlins = 1.35 m.

∴ Live load = 514 × 1.35 = 693.9 N/m.

3. Wind Load

Basic wind velocity near Visakhapatnam = 50 m/sec,

$k_1 = 1.0$

k_2 for category 2, class B building with height 8 m, is 0.98,

$k_3 = 1.0$

∴ Design wind pressure $p_j = 0.6 \times 49^2 = 1440\ N/m^2$

Wind Pressure Coefficients

$$\frac{h}{w} = \frac{10}{15} = \frac{2}{3}$$

Thus,

$$\frac{1}{2} < \frac{h}{w} < \frac{3}{2}.$$

From Table:

When wind angle 0°, for rafter slope 21.8° (wind normal to ridge).

On windward side: $Cpc = -0.7 + \frac{1.8}{10} \times 0.5 = 0.61$

On lee ward sider: $C_{pc} = -0.5$

When wind angle 90°, for rafter slope 21.8° (wind parallel to ridge),

On windward side $C_{pc} = -0.8$

On lee ward side: $Cpc = -0.6 - \frac{1.8}{10} \times 0.2$

$= -0.636.$

Internal Wind Pressure Coefficient

For a building with medium permeability,

$C_{pl} = \pm 0.5$

∴ Design wind pressure on windward side,

$= (-0.8 - 0.5)p_d$

$= -1.3 \times 1440 = -1872 \text{ N/m}^2$

Design wind load on leeward side,

$= (-0.636 - 0.5) \times 1440 \text{ kN/m}$

$= -1635.8 \text{ N/m}^2.$

For Purlin Design

$DL + LL = 364.5 + 694.9 = 1059.4 \text{ N/m}.$

$= 1.0594 \text{ kN/m}.$

WInd Load,

$= -1872\text{N/m}^2 = -1872 \times 1.35 \text{ N/m}.$

$= -2527.2 \text{ N/m}.$

2. Let us discuss the problem based on the purlin for a roof truss of an industrial building located at Chennai with a span of 15m and length of 60 m. The roofing is AC sheeting. The terrain is an open industrial area. Building is class B building with a clear height of 8m at the eaves. Design for 1.5 (DL + WL) combination. Type of truss and roof slope can be assumed.

Solution:

Design of Purlin

Industrial Building - Chennai.

Span = 15m

Length = 60m

Roofing = AC Sheeting

Terrain = Open

Building = Class B

Height = 8m @ Caves

Load Combination = 1.5[DL + WL].

Assume Spacing of truss = 5m C/S

Spacing of Purlin = 1.5 m C/C

Number of Bay = 12 Nos.

Assume Load from roofing material = 200 N/m.

Wind Load = 1200 N/m².

Live Load = 600N/m².

Inclination of main raffer = $\tan\theta = \frac{1}{5}$

$\theta \Rightarrow 11.31°$.

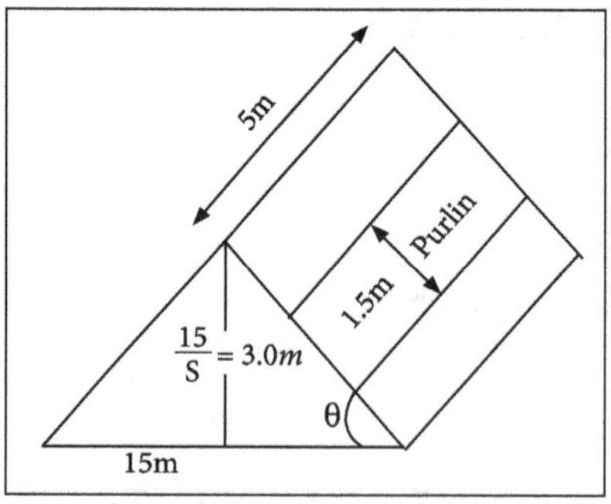

Let us assume a live load of 600 N/m².

Step 1:

$$\text{Minimum depth} = \frac{L}{45} = \frac{5000}{45} = 111 \text{ mm.}$$

$$\text{Minimum width} = \frac{L}{60} = \frac{5000}{60} = 83 \text{ mm.}$$

Hence Let us try ISA 125 × 95 × 8mm angle section.

Step 2: Dead Load and Live Load

D.L = 200 N/m²

= 200 [Spacing of Purlin]/m

= 200 × 1.5

D.L = 300N/m.

L.L = 600 N/m²

= 600[Spacing of Purlin]/m.

= 600 [1.5]

L.L = 900 N/m².

Total Load = 1200 N/m.

W_u = 1800N/m.

Factored Load 1.5 [D.L + L.L] along sheeting

= 1800 cos 11.3°

= 1765 N/m.

Factored Load 1.5 [D.L + L.L] normal to sheeting,

= [450 – 1800]

= – 1.350 N/m.

∴ DL + LL is critical,

∴ It has a greater value of 1765 N/m.

Bending Moment

$$M = \frac{WL^2}{10} = \frac{1765 \times 5^2}{10}$$

= 4412.5N.m.

ISA125 × 95 × 8 mm

$$\frac{b}{t} = \frac{95}{8} = 11.875 \;\&\; \frac{d}{t} = \frac{125}{8} = 15.625$$

Referring Table of IS 800:2007,

b/t 10.5 and 15.7

d/t 10.5 and 15.7.

It belongs to class 3 semi compact section,

$$M_d = \beta_b Z_P f_y / \gamma_{m_0}$$

$$\beta_b = \frac{Z_e}{Z_P}$$

$$M_d = Z_e f_y / \gamma_{m_0}$$

$$= \frac{30.6 \times 10^3 \times 250}{1.1}$$

$$= 6954545 \text{ N} - \text{mm.}$$

$$= 6954.54 \text{N.m} > 44112.5 \text{N.m.}$$

Hence Safe.

3. Let us discuss about an industrial building the trusses of 16 m span and 4 m rise are spaced at 8 m a part. The building is in medium wind zone in an industrial area of plain land.

Solution:

Span = 16 m

Spacing of truth = 8 m.

Assume line load = 600 N/m^2.

Inclination of main rather = tance = 1/4 = 14.03°.

Step 1: Trial Section

Assuming minimum depth $= \frac{L}{45} = \frac{8000}{45} = 177.78\text{mm}$

Minimum width $= \frac{L}{60} = \frac{8000}{66} = 133.33\text{mm}.$

Let us try ISA 125 × 95 × 8 mm angle section.

Step 2: Dead Load and Line Load

Dead load from roofing materials,

$= 200\ N/m^2$

$= 200 \times 1.5$

$= 300\ N/m$

Line load = 600 × 1.5

$= 9\omega\ N/m.$

Factored (DL + LC) along sheering = 1200 × 1.5 × cos (14°)

$= 1746.5\ N/m$

Normal to sheering = (300 – 1200) (1.5)

$= -1350\ N/m.$

D.L + L.L is critical since it has a greater value 1746 N/n.

Step 3: Bending Moment Calculations

Maximum bending moment $M = \frac{WL^2}{10}$

$$= \frac{1746 \times 8^2}{10}$$

$$= 11174.4\ N - m$$

For the chosen ISA 125 × 95 × 8 mm,

$$\frac{b}{t} = \frac{95}{8} = 11.875$$

$$\frac{d}{t} = \frac{125}{8} = 15.625$$

Referring to Table of IS 800:2007,

$$\frac{b}{t} > 10.5 \text{ and } 15.7$$

$$\frac{d}{2} > 10.5 \text{ and } 15.7.$$

$$M_d = \beta_b Z_P \frac{f_g}{\gamma_{mo}}$$

$$\beta_p = \frac{Z_e}{Z_p}$$

$$M_d = \beta_b Z_p \frac{f_y}{\gamma_{mo}}$$

$$= Z_e \cdot \frac{f_y}{\gamma_{mo}}$$

For ISA 125 × 95 × 8 mm

$$Z_e = 30.6 \times 10^3 \text{ mm}^3$$

$$M_3 = \frac{30.6 \times 10^3 (250)}{1.1}$$

$$= 6954545 \text{ N} - \text{mm}$$

$$= 6959.54 \text{ N} - \text{m}$$

$$= 4412.5 \text{ N} - \text{m}.$$

Hence let us provide ISA 125 × 95 × 8 mm for the piston.

4. Let us discuss the problem based on a purlin using the following data:

- Spacing of roof trusses = 4.5 m c/c.
- Purlin spacing = 1.8 m c/c.
- Pitch of roof = 1/4.
- Span of the roof = 10 m.
- The vertical load from roof sheets = 180 N/m2.
- Wind load intensity normal to roof = 1200 N/m2 Use I section.

Solution:

Design a Purlin

Spacing of roof truss = 4.5mc/c.

Purlin Spacing = 1.8 m c/c.

Pitch of Roof = 1/4.

Span of the Roof = 10 m.

Vertical load from roof Sheets = 180N/m^2 .

Wind Load intensity normal to roof = 1200 N/m^2.

Use I section.

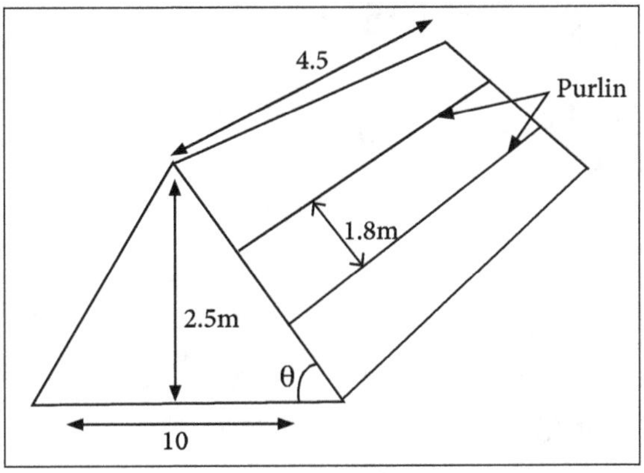

Let us assume a live load of 600 N/m^2.

Inclination of main rafter $\Rightarrow \tan\theta = \dfrac{1}{4}$

$\theta = 14.03°$.

Resolving Forces Parallel and Perpendicular to Sheeting

Dead Load of roofing material = 180 N/m^2.

Spacing of Purlins = 1.8m.

Dead Load = 180 × 1.8 = 324N/m = 0.324kN/m.

Assume Live Load = 1.5 kN/m^2

= 1.5 × 1.8 = 2.7 kN/m

Factored Vertical Load,

= 1.5 [D.L + L.L]

= 4.536 kN/m

Hence Load acting normal to Sheeting,

= 4.54 cos 14.03

= 4.404 kN/m.

Load acting parallel to Sheeting,

= 4.54 sin 14.03

= 1.1 kN/m

Calculating of B.M and S.F about ZZ-axis: B.M

$$M_Z = \frac{W_Z l^2}{8} = \frac{14.404 \times 4.5^2}{8} = 11.4 \text{kN.m}$$

$$M_Y = \frac{W_Y l^2}{8} = \frac{101 \times 4.5^2}{8} = 2.78 \text{kN.m}$$

S.F,

$$F_Z = \frac{11.14 \times 4.5}{2} = 25 \text{kN}$$

$$F_Y = \frac{2.78 \times 4.5}{2} = 6.25 \text{kN.}$$

Choosing of Trial Section

Let us try with I SMB 150 section channel.

Section having following properties.

$r_1 = 9$ mm, $D = 156$ mm, $b = 80$ mm, $t_f = 7.6$ mm, $t_w = 4.8$ mm,

$Z_{ez} = 96.9 \times 10^3$ mm^3.

$Z_{ey} = 13.1 \times 10^3$ mm^3.

$Z_{pz} = 110.5 \times 10^3$ mm^3.

$d = 150 - (7.6 + 9) = 116.8$ mm.

$I_z = 726.4 \times 10^4$ mm^4.

$$Z_{PZ}\text{required} = \frac{M_Z}{f_y}\gamma_{m_o} + 2.5\frac{d}{b}\frac{M_Y}{f_y}\gamma_{m_o}$$

$$= \frac{[11.14]}{250} \times 10^6 \times 10^1 + 2.5\frac{116.8}{80} \times \frac{2.78 \times 10^6}{250} \times 1.1$$

$$= 93662.8 \text{ mm}^3 < 110500 \text{ mm}^3.$$

Hence Section is Okay.

Check for Shear

$$V_{dz} = \frac{f_y / \sqrt{3}}{\gamma_{mo}} \times \text{Area of Web}$$

$$= \frac{250/\sqrt{3}}{101} \times 150 \times 4.8$$

$$= 94478 \text{ N} > 25000\text{N}$$

$$V_{dy} = \frac{f_y / \sqrt{3}}{\gamma_{mo}} \times \text{Area of Flange}$$

$$= \frac{250/\sqrt{3}}{1.1} \times 2 \times b \times t_F$$

$$= \frac{250}{\sqrt{3} \times 1.1} \times 2 \times 80 \times 7.6$$

$$= 159563 \text{ N} > 6250 \text{ N}.$$

Hence the section is adequate in shear.

Calculation of Design Capacity of the Section

$$b/t_f = \frac{80}{7.6} = 10.526 > 9.4\varepsilon < 10.5\varepsilon$$

$$d/t_w = \frac{116.8}{4.8} = 24.33 < 84\varepsilon.$$

By referring Table of IS800-2007 for rolled section with NA @ mid depth.

It belongs to compact section, $\beta_b = 1.0$.

Referring to CI.8.2.1.2 of IS800-2007,

$$M_{d_z} = \beta_b Z_p f_y / \gamma_{mo} < 1.2 Z_e f_y / \gamma_{mo}$$

$$= \frac{1 \times 110.5 \times 10^3 \times 2.5}{1.1} < \frac{1.2 \times 96.9 \times 10^3 \times 250}{1.1}$$

$$M = 25.11 \times 10^6 \text{ N-mm} < 26.43 \times 10^6 \text{N-mm}$$

$$M_{d_y} = \frac{f_y Z_{PY}}{\gamma_{mo}} \left[\because Z_{PY} = b_{tf} \frac{b}{2} \right]$$

$$= 80 \times 7.6 \times \frac{80}{2}$$

$$= 24320 \text{ mm}^3$$

$$= \frac{250 \times 24320}{1.1} = 5.527 \times 10^6 \text{ N} - \text{mm.}$$

Check Whether

$$\frac{M_z}{M_{dz}} + \frac{M_y}{M_{d_y}} \leq 1.0$$

$$\frac{11.14 \times 10^6}{25.11 \times 10^6} + \frac{2.78 \times 10^6}{5.53 \times 10^6} \leq 1.0$$

$$0.94 < 1.0$$

Hence Safe.

Check for Deflection of Purlin

$$I_z = 726.4 \times 10^4 \text{ mm}^4 . L = 4500 \text{ mm}$$

$$\omega = 4.54 \text{ kN/m} = 4.54 \text{ N/mm. } E = 2 \times 10^5 \text{ MPa.}$$

$$\delta = \frac{5}{384} \times \frac{10 L^4}{EI} < \frac{L}{150} [\text{Permissible deflection}]$$

– As per Table – IS800-2007.

$$= \frac{5}{384 \times 4.45 \times 4500^4 < 4500 - 1\lambda} = 16.69 < 30 \text{ mm}$$

5. Let us discuss a channel section purlin for the following data; Spacing of trusses = 4.2 m Spacing of purlins = 2 m.

Solution:

Line load on galvanized iron roofing sheet = 0.6 kN/N².

Wind load =1.4 kN/m^2.

Slope of main rates = 31°.

Step 1: Resolving Forces Parallel and Perpendicular to Sheeting

Assume dead load = 200 N/m^2

Let the self-weight of purlins = 125 N/m^3

Total dead load,

$= 200 + 125$

$= 325$ N/m^2

$= 0.325$ kN/m^2

Live load = 0.6 kN/m^2.

Factored DL + LL is given as,

$= 1.5\ (0.325 + 0.6)$

$= 1.3875$ kN/m^2

Spacing of purlin = 2 m.

Load acting vertically downward,

$= W = (1.3875)\ (2)$

$W = 2.775$ kN/m.

Hence load acting normal to sheeting

$W_x = 2.775 \cos\theta$

$= 2.775 \cos 30°$

$W_x = 2.403$ kN/m.

Load acting parallel to sheeting,

$W_y = 2.775 \sin\theta$

$W_y = 1.3875$ kN/m.

Step 2: Calculation of BM and SF about Z – Z axis

Bending moments,

$$M_z = \frac{l^2}{8}$$

$$M_y = \frac{l^2}{8}$$

$$1 - \text{span of purlin} = 4.2 \text{ m}$$

$$M_z = \frac{(2.403)(4.2)^2}{8} = 5.298 \text{ kN} - \text{m}$$

$$M_y = \frac{(1.3875)(4.2)^2}{8} = 3.059 \text{ kN} - \text{m}$$

Shear Forces,

$$F_z = \frac{(2.403)(4.2)}{2} = 5.0463 \text{ kN}$$

$$F_y = \frac{(1.3575)(4.2)}{2} = 2.91375 \text{ kN.}$$

Step 3: Choosing of total section

Let using ISMC 125 channel section.

$$Z_{ez} = 66.6 \times 10^3 \text{ mm}^3$$

$$Z_{ey} = 13.1 \times 10^3 \text{ mm}^3$$

$$t_f = 8.1 \text{ mm}$$

$$t_w = 5.0 \text{ mm}$$

$$d = 125 - 2(8.1)$$

$$= 108.8 \text{ m.}$$

$$b = 65 \text{ mm}$$

$$Z_{P2} = \frac{m_z}{fy}\gamma_{mo} + 2.5\,\bar{b}\,\frac{M_y}{fy}\gamma_{mo}$$

$$= \frac{5.298 \times 10^6}{250}(1.1) + 2.5\left(\frac{108.8}{65}\right)\left(\frac{3.059 \times 10^6}{250}\right)(1.1)$$

$$= 23311.2 + 56323.2$$

$$= 79634.4 \text{ mm}^3$$

Z_{p2} of ISMC 125 = 79.63×10^3 mm³.

Hence adequate.

Step 4: Check for Shear

$$V_{dz} = \frac{fy/\sqrt{3}}{\gamma_{mo}} \times \text{Area of Web}$$

$$= \left(\frac{250/\sqrt{3}}{1.1}\right)(108.8 \times 5) - 71384 = 71.38 \text{ kV } ZF_2.$$

$$V_{dy} = \frac{fy/\sqrt{3}}{\gamma_{mo}} \times \text{Area of flanger}$$

$$= \frac{250/\sqrt{3}}{1.1}(2 \times 65 \times 8.1)$$

$$= 138174 = 138.17 \text{ kN Fy}.$$

Hence the section is adequate is sheet.

Step 5: Calculation of Design Capacity of the Section

Classification of section,

$$\frac{b}{t_s} = \frac{65}{8.1} = 8.025$$

$$\frac{d}{t_w} = \frac{108.8}{5} = 2.176$$

$$\text{IS800:2007 } \frac{b}{t_5} < 9.4$$

The section classification is plastic,

$$\frac{d}{t_w} < 4.2$$

$$M_{dz} = \beta_b \; Z_p fy / \gamma_{mo}.$$

For plastic section $\beta_b = 1.0$,

$$M_{dz} = \frac{\beta_b \; fy \; Z_{py}}{\gamma_{mo}} = \frac{(1.0)(250)(77.2 \times 10^3)}{1.1}$$

$$= 17545454 \; N-mm$$

$$\approx 17.54 \times 10^6 \, N-mm.$$

$$\frac{1.2 Z_{ez} \, f_y}{\gamma_{mo}} = \frac{1.2(66.6 \times 10^3)(250)}{1.1}$$

$$= 18163636 \; N-mm$$

$$17.54 \times 10^6 \; N-mm(M_z).$$

$$M_{dz} = 17.54 \times 10^6 \, N-mm$$

$$M_{dy} = \frac{fy \, Z_{py}}{\gamma_{mo}}$$

Where,

$$Z_{py} = b \, t_f \; (b/2)$$

$$= \frac{b^2}{2} \; t_f$$

$$= \frac{(65)^2 \, (8.1)}{2} = 17.111 \; mm^3.$$

$$M_{dy} = \frac{250(17111)}{1.1} = 3888920 \; N-mm$$

$$\frac{1.5 \; Z_{zy} \; fy}{\gamma_{mo}} = \frac{1.5(13.1) \times 10^3 (250)}{1.1}$$

$$= 4465909 \text{ N} - \text{mm}.$$

$$= 3888920 \text{ N} - \text{mm}.$$

$$M_{dy} = 3888920$$

$$= 3.889 \times 10^6 \text{ N} - \text{mm}.$$

Step 6: Check Weather

$$\frac{M_z}{M_{dz}} + \frac{M_y}{M_{dy}} \leq 1.0$$

$$M_z = 4.806 \times 10^6 \text{ N} - \text{mm}.$$

$$M_y = 2.775 \times 10^6 \text{ N} - \text{mm}.$$

$$\frac{4.806 \times 10^6}{17.54 \times 10^6} + \frac{2.775 \times 10^6}{3.889 \times 10^6} = 0.9876 < 1.$$

Hence the section is adequate.

Step 7: Check for Deflection

1_z for ISMC 125 = 416.4×10^4 mm^4

W = 2.775 kN/n = 2.775 N/mm.

Actual deflection is given by,

$$\delta = \frac{5 Wl^4}{384 \text{ EI}} = \frac{5}{384} \left(\frac{2.775 \times (4000)^4}{(2 \times 10^5)(416.4 \times 10^4)} \right)$$

$$E = 2 \times 10^5 \text{ MPa}$$

$$= 11.107 \text{ mm}.$$

As per table of IS 800: 2007.

Permissible deflection $= \frac{L}{150}$

$$= \frac{4000}{150} = 2.6467 \text{ mm}$$

$$= 11.107 \text{ mm}.$$

Step 8: Factored DL = (Load Factor) (Total dead load) (Spacing of Listen)

= (1.5) (325) (2.1)

= 1023.75 N/m

= 1.02 kN/m.

Wind load = 1.5 kN/m²

= (1.5) (2)

= 3 kN/m.

Factored wind force = 1.5 (3)

= 4.5 kN/m.

Load normal to sheeting = – 4.5 + factored DL

= – 4.5 + 1.02 cos 30°

= – 3.66 kN/m (outward).

Load parallel to sheeting = 0.975 sin 30°

= 0.5200 kN/m.

$$M_{zz} = \frac{3.66(4)^2}{8} = 7.32 \text{ kN/m}$$

$$M_{yy} = \frac{0.525\,(4)}{8} = 1.09 \text{ kN/m}$$

$$\frac{M_z}{M_{dz}} + \frac{M_y}{M_{dy}} \text{ should be } \leq 1.0$$

$$\frac{7.32 \times 10^6}{7.382 \times 10^6} \text{ should be } \leq 1.0.$$

$$0.9916 + 0.25 = 1.24 > 1.$$

Hence the section may be reversed by selecting next suitable channel section with greater γ_{yy}.

6. Design a purlin based on the purlin for the following specifications:

Span of Truss = 12mc/c

Pitch = 1/5 of span

Spacing of Truss = 5 m c/c

Spacing of purlin = 1.5mc/c

Load from roofing materials etc. = 200 N/m²

Wind load = 1200 N/m²

Use angle section.

Solution:

Consider 1 meters run of the purlin loads,

Dead load of forcing material = 1.5 × 100 = 240 N/m

Dead load of purlin (assumed) = 120 N/m

Total DL = 360 N/m

Wind load acting normal to rag,

= 1200 × 1.5

= 1800 N/m.

Using Angle Section

Total dead load = 360 N/m

Wind load = 1800 N/m

Total load = 1800 + 360 = 2160 N/m.

Maximum bending moment = $M = \frac{2160 \times 3.5^2}{10}$

$M = 2646 \text{ N/m.}$

Section modulus required = $Z = \frac{M}{f} = \frac{26469 \times 10^3}{165}$

$Z = 16036 \text{ mm}^3.$

Minimum depth of purlin $= \frac{\text{Span of purlin}}{60}$

$$= \frac{3.5 \times 10^3}{60} = 58.3 \text{ mm.}$$

$$\text{Minimum width of purlin} = \frac{\text{Span of purlin}}{45}$$

$$= \frac{3.50 \times 10^3}{45}$$

$$= 77.8 \text{ mm.}$$

Provide 90 mm × 60 mm × 10 mm angle having a section modulus of 18600 mm^3.

5.4 Design of Gantry Girder

Load to be considered for the design of gantry guider are:

- Vertical loads from crane.
- Impart loads from crane.
- Longitudinal horizontal face along the crane rail.
- Lateral thrust across the crane rail.

Serviceability Criteria for the Design of Gantry Girders

- Deflection Limit.
- Vibration Limit.
- Durability Consideration.
- Fire Resistance.

Gantry Girder

A gantry girder in most cases is a laterally unsupported beam, subjected to load accompanied to impact. The girder is subjected to unsymmetrical bending, due to various forces transmitted do it. The following are the forces acting on a gantry girder:

- Vertical load transmitted by the crane.
- Lateral force transmitted due to sudden stopping or starting of the crab on the crane.
- Longitudinal forces transmitted due to sudden stopping or starting of the crane.

Steps Involved in Gantry Girder Design

- The maximum bending moment and shear force on the girder considering suitable impact factor has to be worked.
- The horizontal bending moment due to surge load is calculated.
- The vertical and horizontal shear force is calculated.
- A trial section satisfying following guidelines may be selected:
    - The depth of section is chosen approximately [span/12].
    - The width of compression flange may be chosen as [span/30].
    - The moment capacity of vertical loads should be chosen about 40% more than the moment due to vertical load, to resist the combined moment safely.
- The values for Ixx, IYY and Zp for the trial section is calculated.
- The moment capacity of the chosen section may be checked.
- The criteria for buckling resistance may be checked in relevance to Cl.8.2.2 of IS 800:2007.
- The section should be checked for biaxial bending and shear.
- It should also be calculated/ checked for web buckling and web bearing.
- It should be checked for limiting, deflection in accordance with Table of IS 800:2007.
- The connection may be designed for welding.

Problems

1. Let us discuss a gantry girder provided for mill building to carry an electric overhead traveling crane with the following data:

- Span of gantry girder = 8 m.
- Span of crane girder = 18 m.
- Crane capacity = 300 kN.
- Self-weight of crab alone = 30 kN.
- Self-weight of crane = 130 kN.
- Allowable approach of crane hook from the vertical axis to the gantry girder is 1.2 m.
- Distance between center of wheels moving on the gantry girder and supporting the crane girder to 3.5 m.

- Weight of rail section attached on the top of the gantry girder = 0.3 kN/m.

Solution:

Step 1: Combustion of maximum BM and SF

The concentrated loads for maximum BM.

= Weight of trolley + Lifted load by the crank

= 300 + 30

= 330 kN

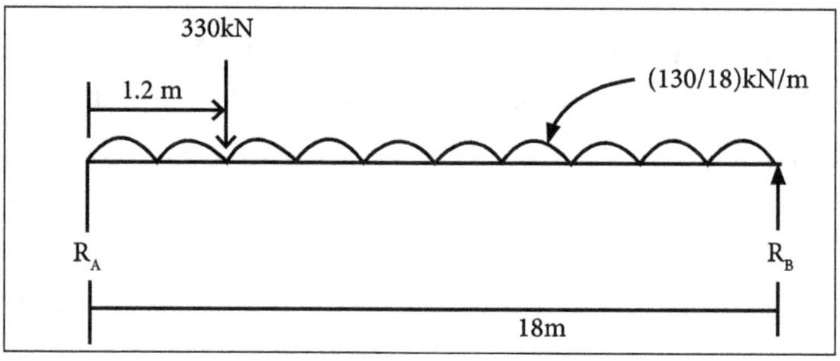

Take moment about B

$$R_A(18)-330\ (18-1.2)-\left(\frac{130}{18}(18)\right)\left(\frac{18}{2}\right)=0$$

$$R_A = 373\ \text{kN}.$$

Load on gantry girder from each wheel,

$$=\frac{373}{2}$$

= 186.5 kN.

Factored wheel load = 186.5 × 1.5 = 279.74 kN.

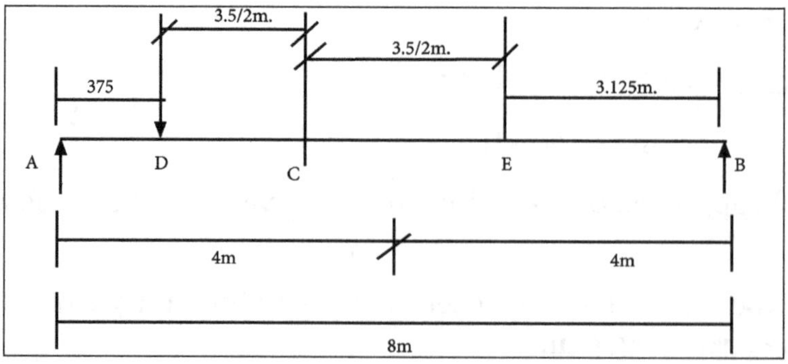

Take Moment About A

$$R_B = 218.5 \text{ kN}$$

$$B_{mat}\ E = R_B\ (3.125)$$

$$= 218.55\ (3 \text{ bits})$$

$$= 638 \text{ kN} - \text{m}.$$

Considering 25% of Impact factor.

Moment due to impact,

= 0.25 (683)

= 170.75 kN-m.

Assuming self-weight of girder as 2 kN/m.

Dead load due to self-weight + weight of rails.

= 2 + 0.3 = 2 – 3 kN/m.

Factored DL = 2 – 3(1.5) = 3.45 kN/m.

$$\text{BM due to DL } = \frac{3.45\,(8)^2}{8} = 27.6 \text{ kN } - \text{ m}.$$

Factored BM due to vertical loads,

= 683 + 170.75 + 27.6

= 881.35 kN-m.

Step 2: Computation of BM Due to Surge Load

Horizontal force transverse to rails,

= 10% of trolley weight and weight lofted

$$= \frac{10}{300}(330)$$

= 33kN.

This load is distributed over and wheel.

Horizontal load on each wheel = 33/4 = 8.25 kN.

Factored horizontal force on each wheel = 1.5 (8.25)

= 12.375 kN.

B_{mat} E = 9.688 (3.125)

= 30.212 kN-m.

Step 3: Computation of Shear Forces

R_A = 437.11 kN.

Vertical shear due to impact $= \frac{25}{100}$ (437.11) × 109.28 kN

Vertical shear due to self-weight of rails = 3.45 (18/2)

= 31.05 kN

Total vertical shear = 437.11 + 109.28 + 31.05

= 577.4 of kN.

R_A= 19.34 kN.

Step 4: Choosing of Trial Section.

Application depth of section $= L/12 = \frac{8000}{12}$

= 667 mm

Width of computer flange $= \frac{800}{2}$ – 267 mm.

Plastic modulus of section = 21602 mm.

Y_P = 510.2 mm

$$Z_P = \frac{M_P}{f_y} = 4810.07 \times 10^3 \text{ mm}^3.$$

For Top Flange

$$Z_{py} = \frac{M_P}{f_y} = \frac{1}{4} \times 21.3 \times 250^2 + \frac{1}{4}(300 - 2 \times 13.6)^2 \times 7.8 + 2 \times 90 \times 13.6 \left(15 - \frac{13.6}{2}\right)$$

$$Z_{py} = 828.485 \times 10s \text{ mm}^3.$$

Check for Moment Capacity

- b/t of flange of ISWB 600 $= \frac{250 - 11.2}{2 \times 21.3} = 5.6 < 9.4\,\varepsilon$

- d/t of web ISWB 600 $= \frac{600-2\times21.3}{11.2} = 49.76 < 83$

- b/t of flange of channel $= \frac{90-7.6}{13.36} = 6.06 < 9.4\,\varepsilon.$

Hence it is a plastic section

$$M_{d2} = 908.8 \text{ kN} - \text{m}.$$

Top Flange

$$M_{d2} = \frac{fy\ Z_P}{1.1} = \frac{250}{1.1} \times 828.48 \times 10^3$$

$$= 188.29 \text{ kN} - \text{m}.$$

$$\frac{1.2 Z_e f_y}{1.1} = \frac{1.2 \times 609.06 \times 10^3 \times 280}{1.1}$$

For top flange $= 166.11 \times 10^6$ N – mm.

As per CI 8.2.1.2 of IS 800.

Check for Combined Local Capacity

$$\frac{638.689}{908.8} + \frac{19.525}{166.11} = 0.82 < 1.$$

Hence adequate.

Check for Buckling Resistance

$$M_d = \beta_d\ Z_P\ f_{bd}.$$

For plastic section $\beta_b = 1$

$$M_d = Z_p b_d$$

$$f_{or} = \frac{1.1\ \pi^2 E}{(L_{LT}/ry)^2}\left[1 + \frac{1}{20}\left(\frac{L_{LT}/ry}{h_f/t_f}\right)^{oy}\right]$$

$$f_{or} = 230.769 \text{MPa}.$$

$$f_{bd} = 145.3 \text{ N/mm}^2$$

$$M_{d2} = 698.9 \text{ kN} - \text{m} < 89.35 \text{ kN}.$$

Let us discuss a bearing stiffener for a welded-plate girder with the following specifications:

- Web = 1000 mm × 6 mm.
- Flange = 2 Nos. of 320 × 20 mm plate on each side.
- Support reaction = 300 kN, and width of support = 300 mm.

Solution:

Outstand of stiffener,

$$= 0.45 \times \text{Width of Flange}$$

$$= 0.45 \times 300 = 135 \text{ mm}$$

Bearing area required $= \dfrac{600 \times 10^3}{180} = 3{,}333.3 \text{ mm}^2$

Thickness of bearing stiffness plate,

$$= \frac{3{,}333.3}{2 \times (135 - 10)} = 13.33 \text{ mm say 15mm.}$$

Total area of the stiffener,

$$= 2(15 \times 135) + (120 \times 6) = 4{,}770 \text{ mm}^2.$$

Induced compressive stress $= \dfrac{600 \times 10^3}{4{,}770} = 125.8 \text{ N/mm}^2.$

Moment of inertia of the stiffener,

$$I = \frac{15 \times 276^3}{12} + \frac{(120 - 15)6^3}{12}$$

$$I = 2{,}628.261 \times 10^4 \text{ mm}^4$$

$$\gamma = \sqrt{\frac{2625.261 \times 10^4}{4770}} = 74.22 \text{ mm.}$$

Slenderness ratio $= \dfrac{l}{r} = \dfrac{0.7 \times 1000}{74.22} = 9.43$

Safe compressive stress corresponding to the slenderness ratio 150 N/mm².

Connections vertical load transmitted to the stiffener per mm height of the stiffener =

$$q1 = \frac{600 \times 10^3}{1000} = 600 \text{ N/mm.}$$

Provide 5 mm wld, strength of the weld per mm length,

$$S_\omega = 0.7 \times 5 \times 110 = 385 \text{ N/mm}$$

$$l_\omega = 50 \text{ mm}$$

$$= \frac{4S_\omega l_\omega}{q_1}$$

Spacing of the weld units $= \dfrac{4 \times 385 \times 50}{600} = 128$ mm.

Clear spacing between weld units,

$= 128 - 50 = 78$mm.

Maximum permissible clear spacing between weld units,

$= 12t = 12 - 6 = 72$mm.

Provide a clear spacing of 70 mm.

Permissions

All chapters in this book are published with permission under the Creative Commons Attribution Share Alike License or equivalent. Every chapter published in this book has been scrutinized by our experts. Their significance has been extensively debated. The topics covered herein carry significant information for a comprehensive understanding. They may even be implemented as practical applications or may be referred to as a beginning point for further studies.

We would like to thank the editorial team for lending their expertise to make the book truly unique. They have played a crucial role in the development of this book. Without their invaluable contributions this book wouldn't have been possible. They have made vital efforts to compile up to date information on the varied aspects of this subject to make this book a valuable addition to the collection of many professionals and students.

This book was conceptualized with the vision of imparting up-to-date and integrated information in this field. To ensure the same, a matchless editorial board was set up. Every individual on the board went through rigorous rounds of assessment to prove their worth. After which they invested a large part of their time researching and compiling the most relevant data for our readers.

The editorial board has been involved in producing this book since its inception. They have spent rigorous hours researching and exploring the diverse topics which have resulted in the successful publishing of this book. They have passed on their knowledge of decades through this book. To expedite this challenging task, the publisher supported the team at every step. A small team of assistant editors was also appointed to further simplify the editing procedure and attain best results for the readers.

Apart from the editorial board, the designing team has also invested a significant amount of their time in understanding the subject and creating the most relevant covers. They scrutinized every image to scout for the most suitable representation of the subject and create an appropriate cover for the book.

The publishing team has been an ardent support to the editorial, designing and production team. Their endless efforts to recruit the best for this project, has resulted in the accomplishment of this book. They are a veteran in the field of academics and their pool of knowledge is as vast as their experience in printing. Their expertise and guidance has proved useful at every step. Their uncompromising quality standards have made this book an exceptional effort. Their encouragement from time to time has been an inspiration for everyone.

The publisher and the editorial board hope that this book will prove to be a valuable piece of knowledge for students, practitioners and scholars across the globe.

Index

www.ingramcontent.com/pod-product-compliance
Lightning Source LLC
LaVergne TN
LVHW080604210926
848485LV00003B/101